SAFETY AND WORKING PROCEDURES IN FOOD ANALYSIS LABORATORY

SAFETY AND WORKING PROCEDURES IN FOOD ANALYSIS LABORATORY

S. N. Jha
Yogesh K.

BIOTECH
2012
BIOTECH BOOKS

ISBN 978-81-7622-253-2

Published by : **BIOTECH BOOKS**
4762-63/23, Ansari Road,
Darya Ganj, New Delhi - 110 002
Phone: +91-011-23262132
E-mail: biotechbooks@yahoo.co.in

Laser Typesetting : **Sushil Kumar**

Printed at : **Chawla Offset Printers**
Delhi - 110 052

PRINTED IN INDIA

Foreword

One of the most demanding tasks for researchers is to follow good laboratory practices (GLP) and standard operating procedures (SOPs) in research organisations to ensure the consistency and reliability of results. The need for proper documentation of GLP and SOP is essential today in any laboratory.

The term 'Good Practice' denotes nothing else than the established way of doing something, the way generally recognised as being the proper one. Standard operating procedures (SOPs) are written procedures for laboratory equipment. They define how to carry out protocol-specified activities.

This book discusses the GLP and SOPs with general safety precautions to be followed in any laboratory in general and Food Quality Analysis in particular. It will also serve as operation manual of general as well as some specific instruments and will pave the way of GLP and SOP in Food Analysis Lab. We hope it will help to a great extent to all researchers in research institutions.

I congratulate to authors and the contributors for this unique contribution for GLP and SOPs to be followed in all laboratories

Dr. M. M. Pandey
Deputy Director General (Engineering)
Indian Council of Agricultural Research, New Delhi

Preface

This manual is intended to be a safety reference document for lab workers in general, It provides basic information about hazards encountered and safety precautions to prevent accidents along with Good Laboratory Practices (GLP) to be followed in any laboratory. It also provides some basic constants, conversion factors, unit of measures as well as some common protocols and equipment standard operating procedures (SOPs).

It also contains the description of faculty, student, and staff roles in ensuring a safe working environment in the lab. It is often a sticky effort to balance convenience and control with regards to safety issues in the lab, but when all is said and done, the basic rules are alertness and vigilance, common sense, conservation of resources, and concern for other workers in the lab.

Any books /documentation cannot be completed without the contributions and suggestions of colleague and peers. Authors acknowledge the contributions of Dr.

K. Narsaiah, Dr. H. S. Oberoi, Dr. S. K. Devatkal, Dr. Pranita Jaiswal, and Dr. Manjunath M. They also express their thanks to RAs and SRFs of NAIP subprojects on "Development of Nondestructive Systems for Evaluation of Microbial and Physico-chemical Quality Parameters of Mango" for helping in compilation of operating procedures of various instruments and equipment.

S. N. Jha
Yogesh Kumar

Contents

Foreword v

Preface vii

1. The Responsibility of Faculty 1
2. The Responsibility of Lab Technician 4
3. The Responsibility of Students/Trainees 7
4. General Safety Precautions 11
5. Safe Methods of Working 15
6. Protective Clothings and Equipment 20
7. Routes of Exposure 26
8. Types of Chemical 30
9. Handling & Storage of Chemicals 36
10. Handling Toxic Materials 39
11. Emergency Protocols 42
12. Chemical Spill 45

13. First Aid 50

14. Laboratory Equipment 54
Akta Prime Fractions Collector 54
Analytical Balance 57
Autoclave ***(Mac Laboratory Equipment)* 60**
Autoclave ***(Pressure Cooker Type)* 62**
Band Sealer 63
Bowl Chopper or Meat Cutter 64
Biochemical Analyzer 66
B.O.D Incubator ***(SANCO)*.......... 68**
Centrifuge 69
Desiccators 73
Digital Infrared Thermometer 73
Digital Refractometer (***PAL-2, Atago; Range: 0-45° Brix; Accuracy: ±0.2%***) 75
Digital Vernier Caliper (***Make: Precision; Range 0-150mm; Accuracy: 0.02mm***) 77
ELISA Reader 78
Environmental Control Chamber ***(SANCO)*.......... 80**
Fermenter ***(30L)* 81**
Flow Injection Analysis System ***(FIA)*** 83
Fluorescence Spectrometer 85
Fume Hood Chamber (***GODREJ***) 87
Gas Cylinders 88
Gas Permeability Tester (***Labthink, Model BTY-B1P, China***) 88

Glass Apparatus 90

Headspace Gas Analyzer/O_2 and CO_2 Analyzer ***(902D Dual Track)*** **91**

Heat Sealer ***(Foot Operated)*** **92**

High Performance Liquid Chromatography ***(HPLC)*** **92**

Hot Air Oven ***(MAC Laboratory Equipment & Others)***.................. **96**

Hunter Lab Colorimeter ***(MINISCAN XE, 45/0-L)***.................. **97**

IR Moisture Analyzer ***(Sartorius, Model MA150)***................. **100**

Laminar Air Flow (***RESCHOLAR***) 102

Lypholiser 103

Microfuge (***Beckman Coulter Centrifuge 22R***). 106

Microscope 107

Mini Lab Fermenter (***2.5 L Capacity***)......... 113

Pipettes 115

Quartz Crystal Microbalance (QCM) ***(QCM200)***........................... **116**

Refrigerators 119

Refrigerated Centrifuge ***(Model MP 400R)*** **120**

Sausage Filler 121

Sirman Meat Grinder 122

Sonicator ***(Sonics Vibra Cell)***............... **124**

Spectrophotometer ***(Dual Beam UV- VIS)***..... **126**

Spectrometer NIR Analyzer AOTF ***(5030 Miniature; Range: 1200-2200nm)* ... 130**

Spectrometer NIR USB ***(EPP2000 NIR InGaAs512; Range: 900-1800nm)* 131**

Spectrometer VIS NIR USB 134

Spectrophotometer UV-VIS (***Model Spectroscan 80DV***) 140

Surface Plasmon Resonance (SPR) 141

Thickness tester (***Model CHY-C2, China; Accuracy: up to 0.1μ***) 143

Ultra Low Temperature Freezer 144

Vacuum Oven 146

Vacuum Packaging Machine ***(Gas Flushing Type)* 147**

Water Bath............................ 148

Water Vapour Permeability Tester/ WVP Tester 150

15. Safe Storage of Laboratory Equipment........ 152

16. Common Chemical Solution/Reagents........ 153

17. Waste Disposal 180

Bibliography 295

1

The Responsibility of Faculty

The faculty is responsible for the administration of the safety program of any laboratory. Every lab in charge should:

1.1 Set a good example by being enthusiastic about safety, and by observing all safety rules and wearing personal protective equipment (e.g. safety glasses, closed-toe shoes and apron).

1.2 Maintain discipline and enforce the laboratory's operational and safety rules, such as:

a. Wearing the prescribed safety gear: safety glasses and aprons;

b. Not eating or drinking inside the lab;

1.3 Orient students in the proper use of the lab facilities and instruments. This includes showing the locations of the fire extinguisher, eyewash, and evacuation plan, and the locations of the valves for the gas, water, and power lines.

1.4 Train students in proper lab techniques, especially those that may pose safety hazards.

1.5 Carefully review all laboratory experiments for possible safety problems.

1.6 Orient the students and researchers on the chemical waste disposal procedures of the laboratory.

1.7 Anticipate and prevent accidents at all times, be alert and vigilant for unsafe conditions, inspect often and intelligently, and take effective corrective action promptly.

1.8 In case of accidents/emergencies, implement proper first aid measures and if necessary coordinate proper and speedy conveyance of the injured person to a medical facility. Notify the parents/guardian of the injured person about the incident.

1.9 Report all accidents. Accidents resulting in medical treatment or observation should be recorded. A formal written report to the competent authority/Head of the Division should be made of all accidents, stating the cause, effect, and recommendations for the prevention of reoccurrence.

1.10 a As **mentor/research supervisor/guide**, the faculty not only guides the student in the direction of his/her research project, but also ensures that the student is not exposed to safety risks in doing this research. The mentor/supervisor should:

a. Orient researchers with the general operations and safety procedures.

b. Discuss with researchers any safety risks involved in the experimental procedures that will be carried out and the materials that will be used in the research.

c. Determine whether researchers need to work continuoously after office hours or overnight, and ensure there are logistical provision for their safety in such cases.

2

The Responsibility of Lab Technician

The lab technician should also be present during the entire laboratory period to assist the faculty/students in managing the experiment and administering the safety procedures of the laboratory. The lab technician should:

2.1 Set a good example by being enthusiastic about safety, and by observing all safety rules (e.g. no eating and drinking in the lab/stockroom) and wearing protective equipment (e.g. eyewear, aprons, etc., where recommended).

2.2 Ensure that safety materials and paraphernalia are well in stock and replenished when used up, and that these are placed in easily accessible locations in the laboratory. These includes fire extinguishers, first aid kits, special containers for chemical and broken glass wastes, and sand for cleaning up spills.

2.3 Manage the stock of chemicals and materials: keep records of materials, supplies, and deliveries. Inform the immediate supervisor well in advance of the need to order supplies before they run out. Ensure that the log book is updated.

2.4 Inspect all safety implements according to the recommended frequency, such as: fire extinguishers, eye wash, first aid kits, and others.

2.5 Provide technical services for laboratories.

a. Prepare chemicals and solutions with accuracy for laboratory work. Prepare chemical unknowns which must be carefully coded and recorded.

b. Under the supervision of faculty member, do various operations such as purification of solvents by distillation.

c. Together with a trained professional, investigate and repair simple electrical circuits and minor electrical breakdown. Make plumbing repairs, mechanical repairs, glassware repairs, and repairs on metal, wood, and plastic.

2.6 Exercise supervision over students with regard to the stockroom transactions and general order in the laboratories.

a. Dispense chemicals, materials, and equipment.

b. Assist the faculty in inspecting the laboratory chemicals/equipment.

c. Ensure that laboratory is provided with properly labelled chemical waste bottles.

2.7 In case of accidents/emergencies, assist the faculty/lab In Charge in implementing proper first aid measures and if necessary assist in the proper and speedy conveyance of the injured person to a medical facility.

2.8 Report all accidents. Accidents resulting in medical treatment or observation should be recorded. A formal written report to the competent authority/Head of the Division should be made of all accidents, stating the cause, effect, and recommendations for the prevention of reoccurrence.

3

The Responsibility of Students/Trainees

The research associates, senior research fellows, junior research fellows, trainees, etc. are responsible for complying with the safety rules and operational procedures of the Lab. Although the laboratory aims to provide a safe working environment for its students, the students should be responsible for their personal safety when they are working in the lab. Every student should:

3.1 Wear proper personal protective equipment:

a. Apron or some suitable protection must be worn at all times in the laboratory. The first thing you do in entering the lab is to put on your apron the last thing on leaving is to remove it.

b. Safety glasses must be worn inside the laboratory wherever applicable. Goggles resting on your forehead are not in a position to protect your eyes. Contact lenses must not be worn.

c. Wear clothing appropriate for work: Do not wear shorts, sandals, or slippers. Wear clothing that would protect your body against contact with chemicals.

d. Confine hair and loose clothing.

3.2 Be informed about experimental procedures and appropriate safety concerns:

a. Come in the lab prepared, having read the experimental procedure in advance. Do not perform unauthorized work, preparations, and experiments. If in doubt, consult with your lab in charge.

b. Research the hazards and physico-chemical properties of the chemicals used in the experiment, e.g. corrosiveness, flammability, reactivity, and toxicity. Follow hazard precautions. Read and prepare for remedies in case of exposure.

c. Read and understand general safety procedures, and those specific to your experiment.

d. Know the location of the fire extinguisher, fire escape, eyewash and first aid kit.

e. Learn how to use instruments properly, how to dispense the correct amount of reagents properly, and how to dispose of chemical waste properly.

3.3 Avoid unnecessary exposure to chemicals:

a. Follow protocols for handling chemicals and for dealing with spills.

b. Never eat, drink, smoke, use medication, or apply cosmetics inside the lab or storage areas.

c. Keep personal belongings, such as bags, books, or cellphones, away from the work area.

d. Never use your handkerchiefs or face/hand towels to wipe things in the lab.

e. Always wash hands, arms, and face before leaving the work area. This should be done even if gloves are used.

3.4 Keep a neat and clean work area. When you are organized, you are less likely to commit accidents.

3.5 Never work alone, especially with hazardous chemicals.

3.6 Consider lab work serious work, and treat all chemicals with respect. No pranks, horseplay, or other acts of mischief.

3.7 Report all accidents, whether major or minor, to the lab in charge, research supervisor, or the lab technician.

4

General Safety Precautions

4.1 Put off shoes/chappals etc. in laboratory where instructed.

4.2 In areas where chemicals are being used, stored or handled, strictly avoid smoking, consuming any eatables and drinks.

4.3 Only approved laboratory procedures and the safe methods laid down in this manual should be followed for carrying out all laboratory tests and experiments.

4.4 Treat all chemical with respect. Some chemicals can injure or even kill if they are not handled properly.

4.5 Read the labels on all the containers before use. Keep the chemicals in original containers until used. Never put chemical into unmarked or mislabeled containers. Even drinking water should not be stored kept in containers used in laboratory.

4.6 Do not distract other's attention and concentrate on your job at all times to prevent accidents.

4.7 Always keep your work area clean and tidy. Objects lying on the ground can be hazardous and even small spills can be slippery hence clean them immediately.

4.8 Test. requiring use of open flame shall be restricted & allowed only in designated location.

4.9 Use hand gloves, apron and eye protections while handling acids & corrosive chemicals *etc.*, to avoid injuries.

4.10 When drawing samples of toxic and corrosive chemicals, use proper protective equipment.

4.11 Chemical fires can start easily and spread quickly hence remove sources of fire sparks from areas where chemicals are stored or used.

4.12 Certain airborne dusts of chemicals can explode sometimes, so avoid all ignition sources in the area while working with such chemicals.

4.13 Handle flammable chemicals with great respects. These chemicals catch fire easily at normal temperatures and the vapours can explode violently by a spark, flame or heat source.

4.14 The vapours of nearly all-flammable chemicals are heavier than air. This means that the vapours can crawl along the ground to distant flames and sparks then flash back to the source.

4.15 Keep all containers of flammable liquids closed when not in use.

4.16 Empty containers are dangerous as they may contain chemical vapours or liquid residual, which can also be explosive. Keep away from all ignition sources.

4.17 All laboratory personnel must be familiar with location and the use of fire extinguishers, eye baths, fire blankets, gas masks and electrical switches.

4.18 Fire safety equipment should not be blocked by placement of materials.

4.19 When there is a spill or leak of chemicals, extinguish all sources of flame or sparks. Shut off the leak if safe to do so.

4.20 If hazardous chemicals are spilled or released, evacuate the area immediately and re-enter only when you have worn the proper protective equipment. Do not operate any electrical switch as this may generate spark.

4.21 Ensure proper ventilation while working in the laboratory.

4.22 Do not depend on your sense of smell to warn you about vapours. Some chemicals can be hazardous in concentration too low to detect by smell. Other can damage your sense of smell.

4.23 Use all protective equipment precisely while handling hazardous chemicals. Before using any respirator check for proper operation.

4.24 Wash your hands thoroughly with soap and water after handling any chemical or containers of chemicals. Always wash before eating, smoking or using toilet facilities.

4.25 Laboratory electrical equipment should be properly earthen and one should ensure that before using the equipment.

4.26 No electrical equipment should be tampered by unauthorised persons.

4.27 Electrically operated equipment must be switched off immediately after use.

4.28 Water tapes should be turned off at the end of the day shift.

5

Safe Methods of Working

5.1 When distilling solvents, hydrocarbons or other liquids, do not fill a distillation flask to more than ¾th of its rated capacity. Use small nonreactive stones or glass beads in distillation flask where bumping may occur.

5.2 Never bring leaking sample containers into laboratory.

5.3 Consider all electric wires with defective insulation as live and dangerous. Do not overload electrical equipment or use for any purpose other than that for which it was designed. When a liquid has been spilled over an electric equipment, first of all turn off the current and then clean the equipment thoroughly and dry. Do not tamper with electrical equipment. Consult the concerned electrical persons when you find any electrical problem.

5.4 It is always important to remember the location of first aid fire fighting equipment, eye wash showers, emergency safety equipments *etc.* in the laboratory.

5.5 Gas burners should not be left unattended. The gas line tubing to be checked for leakages time to time.

5.6 Hot plates should not be kept in 'ON' condition when it is not in use. All flammable and combustible materials should be kept away form hot plates.

5.7 It is always important to keep the working area neat and clean to avoid any kind of untoward incidents because of bad housekeeping.

5.8 Avoid keeping different chemicals in the same container. The practice of keeping laboratory chemicals in a refrigerator along with drinking water should be avoided. Do not keep food items and drinks in refrigerator meant for storing of chemicals.

5.9 At the time of drawing liquid gaseous samples, it is always important to drain the initial hold up by opening the valve very slowly to avoid sudden splashing.

5.10 Before drawing samples, one should ensure that the sample is sufficiently cooled in the cool pot, to avoid cracking of sampling bottles.

5.11 After drawing samples, a label should be put immediately on the sampling bottles.

5.12 The person collecting the sample should wear the personnel protective equipment like hand gloves, face shield etc.

5.13 Only authorized persons should collect the samples.

5.14 Do not try to mix all the remnant chemicals after analysis for disposal. They should be treated individually as per the laid down procedure.

5.15 Chemicals, which are required to be stored in dark area, should not get exposed to sunlight.

5.16 Safety glass/Face shield needs be used while working with corrosive chemicals such as acids and alkalies so that a splash to the eyes can be avoided.

5.17 While cutting glass rods using glasscutters, wear a face shield. Towels may be placed over the rod to protect the finger and the hands before breaking the piece.

5.18 Glassware to be heated by an open flame should be rested on an iron wire mesh with an asbestos center to prevent the flame coming in contact with the glass above the level of the liquid. This will avoid cracking of the vessel.

5.19 Pressure flasks, which are to be heated under temperature control by a water bath or electric oven should be shielded so that in the event, if the flask is ruptured, its contents and fragments of broken glass will be safety deflected.

5.20 Pressure should be released from or applied to glass vessels slowly.

5.21 It is generally unsafe to put pressure on laboratory glassware to expel its contents.

5.22 Fill reagent and other bottles not more than ¾th of their capacity at room temperature leaving one fourth of the capacity as outage to allow for expansion.

5.23 Severe vibration of a vacuum container may result in collapse, equivalent to the shattering effect of an explosion. Vacuum desiccators may be transported in a wooden box.

5.24 Carry beakers and all usual shapes of flasks and bottles with the fingers around the body of the vessel. Do not grasp or hold the volumetric and other long neck flasks at both top and bottom when their contents are being agitated.

5.25 In washing glassware, no more than gentle pressure should be applied in wiping the insides.

5.26 Cracked glassware or glassware with sharp or broken edges should be either fire polished or destroyed.

6

Protective Clothings and Equipment

6.1 Eye and Face Protection

Eye and protective devices must be worn by students, faculty, staff, and visitors in laboratories where chemicals are stored or handled. The type of safety device required will depend on the nature of the hazard and the frequency with which it is encountered. It is the responsibility of the lab in-charge to determine the level of eye protection required and to enforce eye-protection rules.

a. Safety glasses:

Ordinary prescription glasses do not provide adequate protection from injury to the eyes. The minimum acceptable eye protection requires the use of hardened-glass or plastic safety spectacles. These should have: a minimum thickness of 3 mm, impact resistance, passed flammability tests, and lens-retaining frames. Side shields that attached to regular safety spectacles offer some protection from objects that approach from the side but do not provide adequate protection from splashes. Other eye protection should be worn when a significant splash hazard exists. Safety spectacles are recommended for those who require eye protection frequently and/or for long durations (e.g. more than two hours per day). Three dimensions that are important in providing a comfortable fit include: temple length, nose bridge width, and lens diameter.

b. Goggles:

Goggles are not intended for general use. They are intended for wear when there is danger of splashing chemicals or flying particles. For example, goggles should be worn when working with glassware under reduced or elevated pressure and when glass apparatus is used in combustion or other high-temperature operations. Impact protection goggles have screened areas on the sides to provide ventilation and reduce fogging of the lens and do not offer full protection against chemical splashes. Splash goggles ("acid goggles") that have splash -proof sides should be used when protection from harmful chemical splash is needed. There are specific goggles and masks for glassblowing, welding, and intense light sources, such as lasers and ultraviolet light.

c. *Face shields:*

Goggles offer little protection to the face and neck. Full-face shields that protect the face and throat should always be worn when maximum protection from flying particles and harmful liquids is needed. For full protection, safety glasses should be worn with face shields. A face shield or mask may be needed when a vacuum system (which may implode) is used or when a reaction that has a potential for mild explosions is conducted.

d. *Contact lenses:*

Contact lenses do not provide eye protection in the industrial sense, and must be worn only in conjunction with approved safety eyewear. Furthermore, when the work environment entails exposure to chemical fumes, vapor, or splashes, intense heat, molten metals, or highly particulate atmosphere, contact lens use should be restricted. In addition, identification of contact lens wearers should be ensured for appropriate emergency care and for protection in work areas hazardous to the eyes. Only contact lenses worn for therapeutic reasons can be permitted in the laboratory.

6.2 Respiratory protection

Masks may be used to minimize exposure to noxious fumes, but they are by no means total respiratory protection as air masks, which are self-contained breathing apparatuses with their own air tanks. Furthermore, although the wearer is protected against inhalation exposure, he is warned against absorption through the skin of such substances as HCN gas, aniline vapours, etc.

6.3 Use of gloves

Proper gloves should be worn whenever the potential for contact with corrosive or toxic materials and materials of unknown toxicity, sharp edged objects, and very hot or very cold materials. Gloves should be selected on the basis of the material being handled, the particular hazard involved, and their suitability for the operation being conducted. Glove materials are eventually permeated by chemicals. However, they can be used safely for limited time periods if specific use and glove characteristics (i.e. thickness and permeation rate and time) are known. Common glove materials include neoprene, polyvinyl chloride, nitrile, and butyl and natural rubbers. These materials differ in resistance to various substances. Double gloving is recommended when handling highly toxic or carcinogenic materials. Before each use, gloves should be inspected for discoloration, punctures, and tears. Before removal, gloves should be washed, if the material is impermeable to water.

Leather gloves may be used for handling broken glassware, for inserting glass tubes into rubber stoppers, and for similar operations where protection from chemicals is not needed. Insulated gloves should be used when working at temperature extremes. Various synthetic materials such as Nomex and Kevlar can be used briefly up to 538 °C. It is best not to use gloves made either entirely or partly of asbestos, a known carcinogen. It is the responsibility of the lab in charge to determine whether specialized hand protection is needed for any operation and to ensure that needed protection is available.

Clothing and Protective Apparel. The clothing worn by lab workers can be important to their safety. Such personnel should not wear loose, skimpy, or torn clothing and unrestrained long hair. Loose or torn clothing and unrestrained long hair can easily catch fire, dip into chemicals, or become ensnarled in apparatus and moving machinery. Skimpy clothing offers little protection to the skin in the event of chemical splash. If the possibility of chemical contamination exists, personal clothing that will be worn home should be covered by protective apparel. Finger rings can react with chemicals and also should be avoided around equipment that has moving parts.

Laboratory coats and aprons are intended to prevent contact with dirt and the minor chemical splashes or spills encountered in lab-scale work. These, however, are primarily a protection for clothing, and may itself present a hazard (e.g. combustibility) to the wearer. Cotton (pure or 60:40 cotton-polyester), and synthetic materials such as Nomex or Tyvek are satisfactory; rayon and polyesters are not. Coats and aprons do not significantly resist penetration by organic liquids and, if significantly contaminated by them, should be removed immediately.

Plastic or rubber aprons provide better protection from corrosive or irritating liquids but can complicate injuries in the event of fire. Furthermore, plastic aprons can accumulate considerable static charge and should be avoided in areas where flammable solvents could be ignited.

Lab workers should know the appropriate techniques for removing protective apparel, especially any that has become contaminated. Chemical spills on leather clothing or accessories (watchbands, shoes, belts) can be especially hazardous because many chemicals can be absorbed in the leather and then held close to the skin for long periods. Such items must be removed promptly and decontaminated or discarded.

6.4 Foot protection

Shoes should be worn at all times in laboratories or other areas where chemicals are used or stored. Perforated shoes, sandals, or cloth sneakers should not be worn in labs or areas where mechanical work is being done. Safety shoes are used to protect the feet against injuries from heavy falling objects, against crushing by rolling objects, or against lacerations from sharp edges. Safety shoes are required for personnel whose job duties require the lifting, carrying, or moving, etc. of objects weighing more than 15 lbs, which, if dropped, would likely result in a foot or toe injury.

7

Routes of Exposure

7.1 Dermal Contact

One of the most frequent exposure to chemicals is by contact with the skin. Spills and splash can result in overt contamination of the skin. Also, laboratory personnel may unconsciously contaminate themselves when they touch work surfaces, glassware, or equipment which become contaminated during experimental activity. A common result of skin contact is localized irritation or dermatitis. However, a number of materials are absorbed through the skin to produce systemic poisoning. The main portals of entry for chemicals through the skin are the hair follicles, sebaceous glands, sweat glands, and cuts or abrasions of the outer layers of the skin. The follicles and glands are supplied with blood vessels, which facilitate the absorption of chemicals into the body. Chemicals can also gain entrance into the body when contaminated hands touch the mouth, nose, eyes, sores, or cuts.

7.2 Inhalation

Inhalation of toxic vapours, mists, gases, or dusts can produce poisoning by absorption through the mucous membrane of the mouth, throat, and lungs and can seriously damage these tissues, by local action. Inhaled gases or vapours may pass rapidly into the capillaries of the lungs and be carried into the circulatory system. The degree of injury resulting from inhalation of toxic substances depends on the toxicity of the material, its solubility in tissue fluids, its concentration, and the duration of exposure. Inhalation hazards are often associated with gases and volatile chemicals, but solids and non-volatile liquids can also present an inhalation hazard for laboratory personnel.

a. Laboratory chemicals in the form of dusts and particulates can become air borne when transferred from one container to another.
b. Grinding and crushing procedures can also produce aerosols.
c. Splash created from spills and during vigorous shaking and mixing also results in aerosol formation.

Many of the particulates generated during such procedures do not settle out but remain suspended in the air and are carried about by air currents in the room. Some of these particulates are capable of being inhaled and deposited in the respiratory tract. For many operations it is not obvious that an aerosol is being generated and laboratory personnel may not be aware that a hazardous situation exist. Actually, all laboratory operations involving an open vessel will result in the release of an aerosol. Such operations include weighing, stirring, pouring, pipeting, injections with a needle and syringe, and removing caps and stoppers. Alert laboratory personnel will take care not to create unnecessary aerosols.

7.3 Ingestion

Ingestion of toxic materials in the laboratory can also occur when contaminated hands come in contact with food items which are placed in the mouth. Food items and utensils themselves can become contaminated when stored in the laboratory. The practice of mouth pipeting can result in aspiration of toxic materials.

7.4 Injection

Accidents involving needles and syringes can results in injection contamination through the skin. The needle and syringe is one of the most hazardous items used in the laboratory. Also, containers of toxic chemicals may break resulting in hazard from contact with broken contaminated glass.

7.5 Ocular Exposure

The eyes are of particular concern because they are so sensitive to irritants. Ocular exposure can occur via splash or when contaminated hands rub the eyes. Few substances are innocuous in contact with the eyes and a considerable number are capable of causing burns and provide for rapid absorption of many chemicals.

8

Types of Chemical

All chemicals have toxic effects at some dose level for some route of exposure. It is therefore wise to minimize exposure to chemicals. Chemicals can have local or systemic effects. Local toxicity refers to the direct action of chemicals at the point of contact. Systemic toxicity occurs when the chemical agent is absorbed into the bloodstream and distributed throughout the body, affecting one or more organs. Toxic effects are also classified as acute or chronic. Acute effects are observed shortly after exposure. Chronic effects result from long-term exposure or appear after a latency period.

Hazardous chemical means any chemical which is capable of causing harm to people and the environment.

Health hazard means a chemical for which there is statistically significant evidence based on at least one study conducted in accordance with established scientific principles that acute or chronic health effects may occur in exposed personnel. It includes chemicals which are carcinogens, toxic or highly toxic agents, reproductive toxins, irritants, corrosives, sensitizers, hepatotoxins, nephrotoxins, neurotoxins, agents which act on the hematopoietic system and agents which damage the lungs, skin, eyes, or mucous membranes.

8.1 Carcinogen

A chemical agent that causes a malignant disease or increases statistically the risk of cancer, whether by initiating or promoting it. Some of these compounds are fairly common materials used in many laboratories, such as chloroform, carbon tetrachloride, benzene, dioxane, hydrazine, thiourea, and o-toluidine.

8.2 Corrosive

A chemical that causes visible destruction of, or irreversible alterations in, living tissue by chemical action at the site of contact.

8.3 Highly toxic

a. A chemical that has a median lethal dose (LD_{50}) of 50 mg/kg body weight when administered orally to albino rats weighing 200-300 g each;

b. A chemical that has a medial lethal dose (LD_{50}) of 200 mg/kg body weight when administered by continuous contact for 24 hrs (or less if death occurs within 24 hrs) with the bare skin of albino rabbits weighing between 2-3 kg each;

c. A chemical that has a median lethal concentration (LC_{50}) in air of 200 ppm by volume or less of gas or vapours, or 2 mg/L or less of mist, fume, or dust, when administered by continuous inhalation for one hour (or less if death occurs within one hour) to albino rats weighing between 200-300 g each.

8.4 Irritant

A chemical, which is not corrosive, but causes reversible inflammatory effect on living tissue by chemical action at the site of contact.

8.5 Sensitizer

A chemical that causes a substantial proportion of exposed people or animals to develop an allergic reaction in normal tissue after repeated exposure to the chemical.

8.6 Toxic

a. a chemical that has a median lethal dose (LD_{50}) of 50 mg/kg but not more than 500 mg/kg body weight when administered orally to albino rats weighing 200-300 g each;

b. a chemical that has a medial lethal dose (LD_{50}) of 200 mg/kg but not more than 1,000 mg/kg body weight when administered by continuous contact for 24 hrs (or less if death occurs within 24 hrs) with the bare skin of albino rabbits weighing between 2-3 kg each;

c. a chemical that has a median lethal concentration (LC_{50}) in air of 200 ppm but not more than 2,000 ppm by volume or less of gas or vapor, or 2 mg/L or less of mist, fume, or dust, when administered by continuous inhalation for one hour (or less if death occurs within one hour) to albino rats weighing between 200-300 g each.

Targeted organ effects

a. **Hepatotoxins:** chemicals which produces liver damage.

   Signs & symptoms: jaundice, liver enlargement.

   Chemicals: carbon tetrachloride, nitrosamines, benzene

b. **Nephrotoxins:** chemicals which produce kidney damage.

   Signs and symptoms: edema, proteinuria.

   Chemicals: halogenated hydrocarbons, uranium.

c. **Neurotoxins:** chemicals which produce their primary toxic effect on the nervous system.

Signs & symptoms: narcosis, behavioral changes, decrease in motor functions.

Chemicals: mercury, carbon disulfide.

d. **Agents which act on the blood or hematopoietic system:** decreases hemoglobin function, deprive the body tissues of oxygen.

Signs & symptoms: cyanosis, loss of consciousness.

Chemicals: carbon monoxide, cyanides.

e. **Agents which damage the lung:** chemicals which irritate or damage pulmonary tissue.

Signs & symptoms: cough, tightness of chest, shortness of breath.

Chemicals: silica, asbestos.

f. **Reproductive toxins:** chemicals which affect the reproductive capabilities including chromosomal damage (mutations), and effects on fetuses (teratogenesis).

Signs & symptoms: birth defects, sterility.

Chemicals: lead, DBCP.

g. **Cutaneous hazards:** chemicals which affect the dermal layer of the body.

Signs & symptoms: defatting of the skin, rashes, and irritation.

Chemicals: ketones, chlorinated compounds.

h. **Eye hazards:** chemicals which affect the eye or visual capacity.

Signs & symptoms: conjunctivitis, corneal damage.

Chemicals: organic solvents, acids.

9

Handling & Storage of Chemicals

9.1 Acids

Operate under all precautionary measures whenever heating acid or performing reactions, which may liberate acid fumes. When making reagents always add acid to water, keep acid off skin and protect eyes from splattering. Use protective safety appliances during handling acids.

9.2 Alkalies

Alkalies can burn skin, eyes and respiratory tract severely. Wear heavy rubber gloves and face shield to protect against concentrated liquids. Use effective fume removal device or gas mask to protect respiratory tract against alkali, dusts or vapours. Protect skin and eyes when working with alkalies as solids or concentrated solution.

9.3 Flammable Liquids

a. When using flammable, keep only required quantity in the work area for immediate needs.

b. Store extra flammable chemicals in a separate place.

c. Keep all ignition sources away while working with flammable chemicals.

d. Avoid spillage of flammable chemicals and in case there is spillage, clean the area of once and put the waste in a bin in safe area.

9.4 Toxic Chemicals

a. Check up the label properly before you move or open or use the contents in it.

b. Use appropriate protective equipment before handling toxic chemicals.

c. Use only suction bulb (aspirator bulb) for drawing toxic chemicals along with pipe. Do not suck 'Chemicals' with mouth.

d. Hot materials shall not be poured into a sink or drain as the presence of water in the drain might cause violent spattering. All waste toxic materials should be properly disposed off before draining them.

9.5 Storage of Chemicals in Laboratory

a. Chemicals which might react together to give off dangerous fumes or cause fires or explosion on accidental breakage should be stored away from one another.

b. Volatile liquids should not be stored near heat sources or placed in direct sun light.

c. Stock reagents are best stored on open shelves within easy reach.

d. The largest containers may be stored on the bottom of the shelves.

e. The corrosive materials should be deep enough to hold contents of bottles if leakage occurs.

10

Handling Toxic Materials

10.1 Access Control

Access to laboratories which have toxic chemicals should be controlled. The laboratory door should be kept closed while experiments are in progress. This not only protects those people who might otherwise enter the lab, but also reduces interruptions to the lab worker which could lead to accidents.

10.2 Personnel Practices

Personnel should wash their hands immediately after completion of any procedure involving toxic chemicals and when they leave the lab. The use of liquid soap dispensers is recommended.

Eating, drinking, smoking, chewing of gum, application of cosmetics, or storage of utensils, food, or food containers should not be allowed in labs where toxic materials are used. The practice of mouth pipetting should also not be allowed. Mechanical pipeting aids are to be used for all pipeting procedures.

10.3 Decontamination of Work Surfaces

Worksurfacesshouldbeprotectedfromcontamination by using disposable plastic backed absorbent paper or stainless steel trays. Other items and equipment which become contaminated during experimental activity should be decontaminated with the appropriate solvent.

10.4 Minimizing Aerosols

Since all procedures involving an open vessel of liquids or powders generate aerosols, the laboratory worker should develop techniques which will minimize the creation of aerosols:

a. Discharge fluids from pipetts as close as possible to the fluid level of the receiving vessel, or allowing the contents to run down the wall of the receiving vessel. Dropping the contents from a height generates greater aerosol.

b. Rapid mixing of liquids with pipettes by alternate suction and expulsion or forcibly expelling material from a pipette should be avoided.

c. Care should also be taken when discarding contaminated gloves or plastic backed absorbent paper used to cover the work surface so that contamination is not aerosolized in the process. These should be placed in properly labeled waste containers for disposal.

d. Dry sweeping or dry mopping contaminated laboratory floors could reaerosolize contamination. Floors should be cleaned with a wet mop.

11

Emergency Protocols

11.1 Dealing with Emergency

What to do:

a. Alert others in the area of the accident.

b. Try to alleviate the immediate need in the situation without compromising your own safety:

   i. Remove an incapacitated co-worker from the accident area

   ii. Turn off fuel and power lines, use fire extinguisher

   iii. Immediately call for help.

c. If the situation is beyond your control and presents a danger to you, evacuate the area and immediately call for help.

Note: For fires, follow evacuation procedures under Fire Emergency.

11.2 Fire Emergency

What to do:

a. Alert others in the area of the fire.

b. If the fire is small and confined to its origin, use the portable fire extinguisher.

WARNING: Do not attempt to fight a fire that is spreading rapidly or if you are not sure how to operate the fire extinguisher. The fire may block your exit.

c. Close the doors to confine the fire, and evacuate the building.

d. Meet the fire department outside of the building and provide information about any special hazards and other considerations.

11.3 Evacuation Procedures

When alerted of a fire, assume that the emergency is real until you are informed otherwise.

a. Secure potential hazards in your area such as open flames, ovens, small appliances, compressed gasses, etc.

b. Close windows and doors in your area.

c. Leave the building using the nearest stair to reach ground level.

d. If you encounter smoke, stay close to the floor, and crawl if necessary.

e. Do not re-enter the building unless you are given permission to do so.

11.4 Fire Extinguishers.

Most chemical laboratory fire hazards require dry chemical multipurpose extinguishers (ABC-Cease Fire). "Gas" extinguishers, e.g. containing CO_2, offer first defense against flammable liquids or electrical fires.

To use the fire extinguisher, remember:

P - Pull the pin.
A - Aim the nozzle.
S - Squeeze the lever.
S - Sweep from side to side.

12

Chemical Spill

Dropping containers or otherwise spilling chemicals in the open lab can result in exposure to hazardous agents. It is important to minimize the extent and possible effects of exposure to these hazardous chemicals.

12.1 Immediate Action

a. When the toxicity of the spilled material is unknown, treat the spill like a potential health hazard by avoiding exposure and seeking assistance from a trained personnel.

b. Everyone should leave the affected area, closing the door and warning others not to enter the contaminated area.

c. Notify the lab in-charge or lab technician.

d. Avoid skin contact and minimize inhalation.

e. Any contaminated clothing should be removed and containerized. These should be laundered separately from other clothing before reuse.

f. Use, as appropriate, safety shower or eye wash fountain.

g. Exposed skin should be thoroughly washed with soap and water. Continue flushing with water for 15 minutes or more.

h. For chemical splashes on the eyes, a minimum of 20-minute flushing with copious amounts of water is recommended. Check for and remove contact lenses. Rotate the eyeballs so that all surfaces are rinsed. Forcibly hold the eyelids open as necessary.

i. Seek medical attention.

12.2 Spill Clean-Up

a. If the material is not particularly volatile, has a low order of toxicity, not highly corrosive, and there is no fire hazard, proceed with clean-up operations.

b. Wear appropriate personal protective equipment: goggles, gloves, and respiratory protection, especially for volatile and toxic spills.

c. Use sand/soil to contain liquid spills, and if applicable, use an absorbent material that will contain the liquids. (See Spills of Specific Types of Chemicals below).

For small liquid spills (<100 mL), tissue papers, sand, or an absorbent can be used to contain the spill. However, tissue papers are not suitable for cleaning up flammable spills.

d. Do not brush up solid spills since this may produce airborne dusts. Add sand to contain the solid spill. If the material is not reactive to water, you can add water to the spills and mixture to act as dust-suppressant. Use tissue papers and dustpan to collect the spill residue.

e. If a volatile, flammable or toxic material is spilled, warn everyone immediately to extinguish flames, and turn off spark producing equipment. Shut down all equipment and vacate the area until it is decontaminated. Report the incident to faculty or lab technician.

f. Do not leave tissue papers or other materials used to clean up a spill in open trashcans in the work area. Dispose them properly.

g. Ventilate the spill area, when necessary by opening windows or use a fan.

12.3 Spills of Specific Types of Chemicals

a. Acids and bases

Avoid contact with skin. Neutralize acids and bases with solid sodium bicarbonate and citric acid, respectively. A quantity of solid not much greater than the volume of the liquid spilled should be sufficient, even of concentrated reagents.

Mix neutralizing chemical into the spill, adding some water to provide solvent for the neutralizing reaction. Use pH paper to determine whether the acid or base have been neutralized. After neutralizing, mopping should follow. Final rinse is clear water. Rinse mop and bucket.

WARNING: Do not clean-up hydrogen fluoride (hydrofluoric acid, HF) with silica-containing materials such as sand or vermiculite.

b. Mercury

Because of the high toxicity of mercury vapour, spilled mercury should be cleaned-up immediately. Mercury spilled into floor cracks can be made non-volatile by amalgamation with zinc dust, or by adding a mixture of finely powdered sodium thiosulfate (85 g) and powdered ethylene diamine tetra acetic acid (EDTA) (15g). Do not use sulfur to cover mercury since this will just complicate disposal. Contaminated materials used to clean up the spill should also be placed in properly-labeled containers.

12.4 Escape of Noxious Gases

a. Immediate action

i. Open windows, and close off the room from the rest of the building.

ii. In any event, evacuate personnel.

iii. Heavy vapors (MW >>29) will lie along the floor. Very much lighter vapors will rise. Leave the room in either upright or crawling position, accordingly.

WARNING: A mask may protect the wearer against inhalation of noxious gases, but be warned against absorption through the skin of such substances such as HCN gas, aniline vapour, *etc.*, and the burning and poisoning through skin by HF gas, *etc.*

13

First Aid

The situations described herein are the ones of general concern in any laboratory. For more specific concerns, refer to a more extensive first aid manual available elsewhere (*First Aid Manual, 2002*).

13.1 First Aid for Specific Types of Accidents

a. Chemicals in the eyes

A minimum of 20-minute flushing with copious amounts of water is recommended. Check for and remove contact lenses. Rotate the eyeballs so that all surfaces are rinsed. Forcibly hold the eyelids open as necessary.

Take (do not send) the victim to the nearby medical supervisor. Keep victim warm. Watch victim carefully for shock.

b. Chemicals on the body

Flush areas copiously with water under safety shower or under the sink. Quickly remove chemically soaked clothing, cutting or tearing them if necessary to get them off rapidly. Modesty should not deter removal of clothing in a true emergency. Water-insoluble oils must be washed off with soap and water.

☠ **WARNING: Do not use solvents to wash off offending oily chemicals. Solvents may cause oil to dissolve in the skin, thus increasing the damage to the skin and promoting poisoning through the skin.**

Follow water-washing with the correct neutralizing solution, as 3% boric acid (remedy for base) or 5% bicarbonate (remedy for acid). Keep moist with neutralizing solution soaked on gauzes.

☠ **WARNING: Do not apply greasy burn ointment to chemical burns; it seals the chemicals under the greasy coating.**

c. Fire burns

Rush victim under safety shower or prevent fire on victim with a blanket (e.g. lab coat, jacket, *etc.*). Burned areas should be cooled immediately, hence the advantage of using the safety shower. Soak affected part in ice water for 20 minutes to reduce deep burning.

☠ **WARNING: Application of cold to large areas of the body produces shock. Treatment with ice water is not recommended for large-area burns. The cold shower will suffice.**

Cover burned areas in 3% boric acid. Keep moist. Only very minor burns should then receive burn ointment. On most fire burns, do not apply burn ointment.

d. Chemical poisoning

Get poison remedies/antidotes from the lab technician. Remedies include: egg albumin, magnesium sulfate, mustard powder, charcoal, and tannic acid, *etc*.

13.2 Transporting the Injured Person

No one should be sent to seek medical attention without accompaniment. The faculty/lab technician or a co-worker must accompany the injured person in seeking medical treatment. There is always the possibility that they may collapse en route, even though at the time you dismiss them they assure you they feel fine and are quite ready to go alone. Also, a victim should not drive a vehicle for several hours after a severe injury.

13.3 Signs of Danger

Watch the victim for blood leaving the head (face turning white) or excess blood in the head (face turning red), heavy sweating, cold sweat, shivering and chills, very rapid pulse, weakness of the arms and legs, pounding heart, shallow rapid breathing, jittery talk, extreme dullness or sluggishness, nausea, vomiting, and difficulty in breathing. All these are signs of trouble and require immediate medical attention.

13.4 Shock

Accident victims experience shock. It may be the most dangerous result of an accident. Shock can be fatal, be on the lookout for it.

a. Symptoms of shock include: cold sweat, extreme paleness of the face, nausea, trembling, and shallow breathing.

b. To restore blood to the head, have shock victim lower head between knees, or better, have him lie with head lower than the body.

14

Laboratory Equipment

The standard operating procedures (SOPs) of different equipment are described in this chapter. In the logbook of particular equipment indicate the date, time and name *etc.* accordingly. It is the responsibility of the user to immediately report any damage or malfunction of the equipment to the lab-in-charge and leave a note on the equipment describing the problem. Also the problem should be entered in log book of the equipment.

Note: In all cases, the manufacturer's manual should be consulted for specific instructions.

Akta Prime Fractions Collector

Purpose

The fraction collector is used for partial purification of proteins (enzymes) for use in different applications. Depending upon the requirement, certain level of purification and concentration of protein/enzyme can be achieved using different columns and methods. The Akta prime comprises a control system, pump, fraction collector and chart recorder, together with valves for buffer selection, sample injection, gradient formation and flow diversion.

Procedure

a. Turn on the system using the main switch at the rear panel. The system now performs a self test, all parameters are automatically set to default value.

b. Turn on the computer. Click prime view icon on the desktop to open prime view module.

c. Use high purity water for the preparation of buffers, samples and chemicals.

d. Filter the buffer and sample through 0.45 µm filters.

e. Put inlet tubing in A1 in the binding buffer and tubing B in elution buffer.

f. Put the three waste capillaries (Brown color) from port 4 & 5 on the injection valve and port on the fraction collector valve in waste.

g. Connect the HisTrap HP 1ml column b/w port 1 on the injection valve and the upper port of the UV flow cell.

h. Fill the fraction collector rack with 18 mm diameter tubes. Adjust the height of the delivery arm using the lock. The tube should always be below the horizontal mark on the tube sensor.

i. Check that the sensor is in the correct position for the tube size. The eluent tubing should be over the centre of the collection tube. Rotate the rack manually until the rear half of the tube sensor rests against tube 1.

j. Press feed tube on the front panel the bowl moves to the correct position to collect the first fraction in tube 1. Make sure that the drop synchronization is turned on.

k. Check the UV lamp filter position and calibrate the pH electrode.

Operation

l. While running application templates, the buffer inlet tubing will automatically be filled with buffer.

m. Connect a sample loop between port 2 and 6 on the injection valve. Make sure that the sample loop is large enough for your sample.

n. Select template, application template, His Tag purification HisTrap one after the other in the main menu and press ok.

o. Set the sample volume and press ok to start the purification run.

p. To finish the run press ok at the method complete prompt.

Cleaning

q. Fill a syringe with 5 times the sample loop volume of deionized water. Rinse the sample loop by injecting the water through injection valve.

r. Put all used inlet tubings in water.

s. In the template menu, select application template and then system wash method.

Precautions

a. The system can be used for most applications only after 15 minutes of lamp warm up.

b. Do not use buffer with high salt concentration.

c. During storage, always fill flow path with 20% ethanol.

d. Drop synchronization cannot be used at flow rate above 3ml per minute.

Analytical Balance

An analytical balance is used to measure mass to a very high degree of precision and accuracy. The measuring pan(s) of a high precision (0.1 mg or better) analytical balance are inside a transparent enclosure with doors so that dust does not collect and so any air currents in the room do not affect the balance's operation.

Once the test sample is put on the analytical balance pan (can be on a weighing paper), the load cell is displaced from its original position; and the coil is displaced by the downward force, which causes the load cell to generate a current and return a compensation circuit. This current is then converted to a voltage and uses electronic circuitry and appropriate software to accomplish proper calibration and show the measured weight. Choose a balance appropriate for the range of the object to be weighed and the precision of a particular analysis

Procedure

a. Prior to use, the balance should be checked for spilled substances and spilled solids should be brushed away.

b. Ensure that the balance is levelled, on a stable surface.

c. Place the balance away from air currents.

d. The pan of balance should be checked to ensure that it is properly seated on the pan mechanism.

e. Turn on the balance. If an error message is received in lieu of the tare value, notify your lab-in-charge.

f. Press the tare button and wait until a stable zero point reading is obtained. Re-press the tare button if necessary.

g. Select the range by pressing the control bar.

h. Releasing and then briefly pressing the control bar again, allow you to switch between the different range.

i. When the desired weighing range has been selected, wait: Display "----", then 0.0000. The balance is now in the weighing mode.

j. Place the object to be weighed on the pan, in the centre area. Allow the reading to stabilize and record the weight. If a container or weighing paper is to be used, press the tare button and wait until a stable zero point reading is obtained. Add the substance to be weighed, allow the reading to stabilize, and record the weight.

Precautions

a. Allow samples/containers to reach room temperature before weighing. Hot samples will generate an upward convection of warm air that will cause an inaccurate reading.

b. Samples that are extremely hygroscopic or lose water rapidly must be weighed in a closed container. Volatile liquids must be weighed in a closed container or with a trapping solvent.

c. "Lumps" are often caused by moisture or electrostatic charge and may not constitute a representative sample.

d. When weighing a sample or reagent, discard excess material that has been removed from the original container. Do not return material to its original container.

e. Do not use regular paper for weighing. Use weighing boats or an appropriate container.

f. Do not weigh containers containing a magnetic stir bar. The magnetic field generated may cause unpredictable effects on the electronic balance.

g. Fingerprints may cause an inaccurate value. Make sure hands are clean and dry and do not contribute to the weighing.

h. Be sure to use a clean and dry spatula or other transfer device to avoid contamination of the article to be weighed

i. Do not weigh objects that can carry an electrostatic charge, such as some plastics. Charged materials can result in unpredictable balance behavior and weighing.

j. Turn off balance. Wash and replace spatula in the rack. Clean up any rubbish or spillages near or on balance, close balance doors.

Autoclave *(Mac Laboratory Equipment)*

It works under the principle of steam with pressure sterilization. Reliable sterilization with moist heat requires temperatures above that of boiling water. These high temperatures are most commonly achieved by steam under pressure in an autoclave. Autoclaving is the preferred method of sterilization, unless the material to be sterilized can be damaged by heat or moisture.

A basic principle of chemistry is that when the pressure of a gas increases, the temperature of the gas increase proportionally. For example, when free flowing steam at a temperature of 100°C is placed under a pressure of 1 atmosphere above sea level pressure *i.e.*, about 15 pounds of pressure per square inch (Psi) and the temperature rises to 121°C. Increasing the pressure to 20 psi raises the temperature to 126°C. In this way steam is a gas, increasing its pressure in a closed system increases its temperature. As the water molecules in steam become more energized, their penetration increases substantially. This principle is used to reduce cooking time in the home pressure cooker and to reduce sterilizing time in the autoclave. It is important to note that the sterilizing agent is the moist heat, not the pressure.

Procedure

a. Fill water in the equipment and check the water level so heating element must be dipped in water.

b. Stack the subject matter to be sterilized into stainless steel basket.

c. Close the lid tightly with the help of locking mechanism.

d. Switch on instrument by main and then long press the start button.

e. Set the sterilization parameters.
   i. Temperature-121°C
   ii. Pressure-15 psi
   iii. Hold on time-15-20 min.

Precaution

a. Heating element must be dipped in water.
b. Avoid overloading of material to be sterilized.

Autoclave *(Pressure Cooker Type)*

Procedure

a. Pressure cooker can also be used for sterilization purpose.
b. The standard autoclave cycle used is 15 psi for 15 minutes.
c. A heat sensitive indicator (autoclave tape) should be used in every load.
d. To prevent accidents, wait until the pressure drops to ~5 psi before opening the exhaust valve. The lid should only be open when the pressure is zero.

Precautions

a. Always wear safety glasses when opening the pressure cooker.

b. Always use heat safety gloves when removing items from the pressure cooker.

Band Sealer

Ideal for heat sealing of laminated or non-laminated pre-made pouches. This machine is used for sealing various pre-made pouches and is a best sealing equipment to be widely used for food & groceries, seed processing units, *etc*.

Procedure

a. Check for proper gas cylinder connection.

b. Switch the band sealer 'ON'.

c. Set the speed of belt conveyor and sealing head. Speed depends on type of material being used for pouch.

d. Set the temperature count between 165 and 190°C.

e. Make sure cooling, sealer and seal heating button are in 'ON' mode.

f. Switch the inflator button 'ON' just before packaging to inflate the gas.

Precautions

a. Use the gas just before the processes starts and stop the gas flow as soon as the process completes.

b. Make sure that proper connection of cylinder and working condition of regulator before use.

Bowl Chopper or Meat Cutter

The bowl cutter is the commonly used meat chopping equipment designed to produce small or very small ("finely comminuted") lean meat and fat particles. Bowl cutters consist of a horizontally revolving bowl and a set of curved knives rotating vertically on a horizontal axle at high speeds of up to 5,000 rpm. Many types and sizes exist with bowl volumes ranging from 10 to 2000 litres. The most useful size for small to medium size processing is 20 to 60 litres. In bigger models bowl and knife speed can be regulated by changing gears. Bowl cutters are equipped with a strong cover. This lid protects against accidents and its design plays a crucial role in the efficiency of the chopping process by routing the mixture flow. Number, shape, arrangement, and speed of knives are the main factors determining the performance of the cutter. Bowl cutters should be equipped with a thermometer displaying the temperature of the meat mixture in the bowl during chopping.

Modern large scale bowl cutters may have devices to operate under a vacuum, which helps to improve colour and texture of the meat products by keeping oxygen out of the meat mixes and avoid air pockets. Cutter knives should be adjusted to a distance of 1-2 mm from the bowl for optimal cutting (check the manufacturers recommendations for each model). Most of the large and high-speed bowl cutters are equipped with mechanical discharger devices for emptying the cutter. The process of chopping in a bowl cutter is used for producing fine comminuted products such as frankfurters, bologna, liver sausage etc., and enables processors to offer a much wider range of products.

Procedure

a. Open the lock of bowl chopper.

b. Lift the bowl cover.

c. Place meat and ingredients.

d. Lock the bowl chopper and bring the knob to 1 or 2 mark.

e. Switch on the main power plug.

f. Push green button to run the chopper.

g. Push the red button to stop the running of bowl chopper.

h. Put off the main.

i. Take out the material from bowl chopper.

j. Use dry paper or cloth to remove all the meat and ingredients attached to the bowl, its cover and base.

k. Soak a sponge duster in liquid detergent and use it to clean the bowl chopper.

l. Clean the chopper with tap water using a wet cloth.

m. Finally use wet cloth dipped in hot water for final cleaning of bowl chopper.

n. Dry the chopper with a clean dry cloth.

Biochemical Analyzer

Biochemical analyzer is used for precise and accurate determination of ethanol concentration in the fermentation broth.

Procedure

a. Before initialization of the instrument make sure that the bottle contains the calibrant (ethanol standard) and buffer.

b. Clean the electrode probe first with 70% isopropyl alcohol followed by cleaning using distilled water.

c. Put 3-4 drops of NaCl solution on the top of dried probe and place the membrane carefully on the top of the probe.

d. Connect the unit to the power supply using the power cord provided with the instrument.

e. Main screen displays [Run], [Standby], [Menu].

f. Select menu and screen display 1) Service 2) Set up 3) Diagnostics.

g. Select Diagnostic, five sub menu options are displayed 1) Motor 2) Pump 3) Probe 4) I/O 5) Sensor.

h. Select probe, it shows ethanol delta (P) or current.

i. For starting the set up wait until current comes down below 5 mA.

j. After every 10 min press option no. 1 (flush), to stabilize the current.

k. New membrane will take 5-12 hrs for proper stabilization.

l. After current stabilization, put the instrument on Calibration mode.

m. Go back to run mode and press option sample and place your sample at port no.1.

n. Reading is displayed on the screen

o. Since the membranes are expensive and can be used for 3-4 days only, a large number of samples should be prepared well in advance and stored refrigerated. They should be analyzed during the period the membranes are in working condition.

Precautions

a. Current should be below 5mA before running the sample.

b. Calibration and wash bottle should be filled before operating system.

c. The probe should be clean and dry before using the membrane.

d. Put 3-4 drops of NaCl solution on the top of dried probe and place the membrane carefully on the top of the probe.

e. Ensure that the membranes are stored properly under the refrigerated conditions as specified by the manufacturers.

B.O.D Incubator *(SANCO)*

It is used for providing optimum temperature for growth of micro-organisms. A BOD incubator is an incubator designed to maintain 20°C necessary to perform a test called Biochemical Oxygen Demand (BOD). It involves incubating samples saturated with oxygen at 20°C for (usually) five days. Such an incubator has a compressor to depress the temperature below ambient and a heater to bring it back up to 20°C. These work against each other and produce very precise temperature control, often as close as +/-0.1°C. The term “incubator” is a general term which includes BOD incubators and other incubators designed to maintain other temperatures. The term used unqualified usually refers to an incubator designed to maintain temperatures of 5°C or more above ambient to as high as 100°C. Incubators designed to maintain temperatures below ambient to as low as about 10°C are generally called low-temperature incubators.

Procedure

a. Switch on the instrument by main.

b. Control panel is provided with ON/OFF switch, indicator, digital controller *etc.*

c. Set the temperature range as desired with help of controlled digital temperature cum controller.

d. Temperature is maintained with accuracy of ±0.1°C.

e. Air is circulated by two motors connected to balanced blowers which operate alternatively and continuously with heating and cooling.

f. Interior illumination is done by three fluorescent tubes provided with automatic 0-24 h timer for cyclic illumination.

Precaution

a. Never use the instrument without stabilizer.

Centrifuge

A centrifuge is a piece of equipment, generally driven by an electric motor (some older models were spun by hand), that puts an object in rotation around a fixed axis, applying a force perpendicular to the axis. The centrifuge works using the sedimentation principle, where the centripetal acceleration causes more dense substances to separate out along the radial direction (the bottom of the tube). By the same token, lighter objects will tend to move to the top (of the tube; in the rotating picture, move to the centre).

Protocols for centrifugation typically specify the amount of acceleration to be applied to the sample, rather than specifying a rotational speed such as revolutions per minute. This distinction is important because two rotors with different diameters running at the same rotational speed will subject samples to different accelerations. During circular motion the acceleration is the product of the radius and the square of the angular velocity ω, and the acceleration relative to "*g*" is traditionally named "relative centrifugal force" (RCF). The acceleration is measured in multiples of "*g*" (or × "*g*"), the standard acceleration due to gravity at the Earth's surface, a dimensionless quantity given by the expression:

$$RCF = \frac{rw^2}{g} = \frac{r(2\pi N)^2}{g}$$

where

g is earth's gravitational acceleration,

r is the rotational radius,

N is the rotational speed, measured in revolutions per unit of time. In physics, this is usually denoted *f* and is called the frequency of revolution.

This relationship may be written as

$$RCF = 1.11824396 \times 10^{-5} r_{cm} N^2_{RPM}$$

where

r_{cm} is the rotational radius measured in centimetres (cm), that is, $r = r_{cm}\ cm$,
N_{RPM} is rotational speed measured in revolutions per minute (RPM), that is,

$N = N_{RPM} RPM = N_{RPM} \times 120\pi\ rad/s$, and

g=(980.665 cm/s^2)(3600$s^2/minute^2$)=3530394 $cm/minute^2$

Precautions

a. The instrument must be positioned exactly horizontally to prevent the instrument moving away from its place during centrifugation.

b. Check if the rubber buffers/cushions are in the buckets.

c. Balance the tubes and buckets in the centrifuge.

d. Turn the speed control slowly to the required speed. Do not use the centrifuge at a higher than necessary.

e. Daily cleaning of the centrifuge buckets with a non corrosive disinfectant is recommended. The buckets should be inverted to drain dry.

f. After any sample spillage wipe and disinfect immediately.

g. When tube breakage occurs remove the broken pieces using forceps and thick gloves and dispose appropriately. The portions of the centrifuge in contact with the blood or potentially infectious agent must be immediately decontaminated. The centrifuge bowl should be cleaned with a disinfectant, and rotor heads and buckets should be autoclaved.

h. Stop the centrifuge immediately if it develops an abnormal sound during operation.

i. Preventive maintenance should be carried out by trained staff, on centrifuge speed using a tachometer every 6 months. The measured and rated speed should not differ by more than 5% under specified conditions. The accuracy of the centrifuge timer should be checked every 6 months. The temperature of the centrifuge should be checked at least monthly under standardized conditions. The agreement between the measured and expected temperature should be within 2°C. Check brushes and bearings every 3 months. Replace if necessary. Check for corrosion and repaint if necessary.

j. All maintenance function checks must be recorded, and all corrective actions documented. Always follow the manufacturer's instructions forinstallation,operation,cleaning,maintenance, lubrication, repair and replacement.

Desiccators

A desiccator is a laboratory apparatus that provides a dry atmosphere. A desiccant, typically anhydrous calcium chloride, absorbs the water vapour from within the enclosure of the desiccator. The anhydrous calcium chloride forms $CaCl_2.2H_2O$; the hydrated water molecules can be easily removed with heat and the calcium chloride can be recycled for subsequent use in the desiccator.

Keep the ground glass surfaces of desiccators greased. Seal and unseal desiccators by a siding action of the lids. Never seal desiccators until its contents have cooled to a reasonable temperature.

Digital Infrared Thermometer

(Fluke; Range: -40 to 650°C; Accuracy: ±1°C)

Infrared radiation is a part of sunlight which can be decomposed reflecting it through a prism. This radiation has energy. At the beginning of the 20th century, Scientifics Planck, Stefan, Boltzmann, Wien and Kirchhoff defined the electromagnetic spectrum activities and established comparations to describe infrared energy.

This makes possible to define energy in relation to wavelengths emitted by a black body. Objects with a temperature over absolute zero can irradiate energy.

The amount of energy grows in proportion to the fourth magnitude of the temperature.

This concept is the Basic principle in the measurement of the temperature by means of infrared. With the emission factor a variable is introduced. The emission factor is a measure for the relationship between radiations emitted by both a grey and a black body at the same temperature. A grey body is an object whose wavelengths have the same emission factor. The emission factor of a grey body does not change with the wavelength, as for instance Aluminium. In general, the emission factor is the same as the absorption factor.

For bright surfaces, the emission factor can be automatically or manually adjusted to correct measurement errors in infrared thermometers. When the emission factor is not constant two or more wavelengths should be measured to solve the problem.

Procedure

a. Switch on the thermometer by pressing the power button.

b. Now press the button near to your first finger. A beam of laser appears.

c. Target the sample with the laser beam and hold it for few seconds.

d. Note the readings directly from the display panel in °C.

Precautions

a. Do not target the sensitive organs like eye with beam of light.

Digital Refractometer (*PAL-2, Atago; Range: 0-45° Brix; Accuracy: ±0.2%*)

A refractometer determines the concentration of a solution by measuring the optical refractive index. Ernst Abbe is credited with having invented the refractometer in 1874 when he published a description of an apparatus for determining the refractive index of solid and liquid substances.

Refractive Index (R.I.)

The Refractive Index of a transparent isotropic medium may be loosely defined as the "bending" power of the medium for a ray of light obliquely incident on its surface. This bending power is expressed precisely in the form of an equation

(Snell's Law). sin i/sin r = Nv where i and r are the angles of incidence and refraction of a beam passing obliquely from a vacuum into the plane surface of the medium. For convenience, it is almost universal practice to refer to air as a standard for refractive indices rather than to vacuum. The refractive index relative to air is denoted by n and is related to Nv by the equation Nv/n = Nair = 1,00027

More exactly, the R.I. is defined the following way: Light travels at different speeds in different media. The denser a medium is, the slower the speed of light in that medium. When light passes from one medium to another at any angle other than 90° it not only changes speed, it also changes direction at the boundary between the two media.

The refractive index (symbolized as n or nD) of a medium is defined as the speed of light in air divided by the speed of light in the medium.

Procedure

a. Clean the prism surface with distilled or tap water.

b. Place approximately 0.3 ml of distilled water on the prism surface and set the zero value.

c. Now place the sample on the prism surface and press the measure key to display the reading.

d. Directly note the reading from the display panel in °Brix.

Precautions

a. Always clean the prism before each measurement.

b. Always set zero using distilled water.

Digital Vernier Caliper (*Make: Precision; Range 0-150mm; Accuracy: 0.02mm*)

The vernier caliper is used in length measurements to gain an additional digit of accuracy compared to a simple ruler.

Procedure

a. Completely close the two ends of the measuring arms by manual dragging. Set the zero of the caliper by pressing the "set zero" button.

b. Firmly handle the sample with left hand. Open the measuring arms of the caliper by gradual dragging so that it fits the sample easily.

c. Keep the sample between two arms of the caliper and gently drag the arms towards the sample so that it just touches the sample without destructing it.

d. See the display panel of the caliper and directly note the readings in mm.

Precautions

a. Set zero before each measurement.

c. Make sure that the measuring arms fits the sample.

ELISA Reader

ELISA reader is used for carrying out specialized enzyme assays such as those for β-glucosidase, β-xylosidase, xylan esterase, which require low enzyme volumes and require expensive substrates. Enzyme-linked Immunosorbent Assays (ELISAs) combine the specificity of antibodies with the sensitivity of simple enzyme assays, by using antibodies or antigens coupled to an easily-assayed enzyme. ELISAs can provide a useful measurement of antigen or antibody concentration. There are two main variations on this method: The ELISA can be used to detect the presence of antigens that are recognized by an antibody or it can be used to test for antibodies that recognize an antigen.

Procedure

a. The program is based on session. A session contains:
    i. A plate template
    ii. A measurement/dispensing area
    iii. A plate layout
    iv. Steps
    v. Calculation information
b. All defined by the user.
c. When a new session is created, the type of plate template is requested. This template selection defines the dimensions of the plate & wells.

The user can enter the well specification with the plate layout. Layouts can be saved as files and layouts can be loaded from files to the session. In the area definition window, the user can select the wells to be measured.

Steps

d. Steps are a collection of basic operations that are performed in a certain sequence. You can designate each step with a specific name.

The steps are: Shake 1, Incubate, Shake 2

Shake 1

e. The parameters of the shake step are:- a) Total time: The duration time of the total shake step is to be given. b) ON time: The ON time of shaking period. The ON time must always be given so that the instrument will shake. c) OFF time: The OFF time of shaking period is to be given. d) Speed (rpm): The speed is defined by the user.

Incubate

f. Here incubation time is specified.

Shake 2

g. Here, the total time of shake, ON and OFF time of shaking period as well the speed of shake is defined by the user.

Measure

h. This step measures the given area strip according to the step parameters.

Executing the session

i. You can run the session by clicking the START tool button.

j. Save the results in the required file.

Environmental Control Chamber *(SANCO)*

It is used for providing optimum conditions (Temperature and Relative humidity) for experiments. It provides with humidifier and dehumidifier facility controlled by *on/off* switches. It is also provided with fluorescent tubes, UV tube with PAR lamp with their controlling switches *i.e.* two for fluorescent tubes, one for UV tube and three for PAR lamp. It provides temperature range 0-110°C on thermostat knob.

Procedure

a. Switch on instrument by main.

b. Set the temperature and humidity range as desired with help of controlled digital temperature cum humidity controller.

c. Fill water in the equipment and check the water level by pressing the red color button and it must be above the marked level.

Precautions

a. Never use the instrument without stabilizer.

b. Always check the water level before humidifier switch on.

Fermenter *(30L)*

The fermenter is used for production of value added compounds such as ethanol, enzymes and organic acids through fermentation.

Procedure

a. Fill the fermenter vessel with medium by removing the top plate.

b. Fill the reagent bottle with alkali and acid solutions of required concentrations.

c. Fit the reagent bottle with a small air filter on one metal plate and fit a length of silicon tubing to other metal plate.

d. Connect a length of silicon tubing to the peristaltic pump head and clamp the tubing securely from the pump head to the two way inlet.

e. The pH electrode should be connected to base unit. Switch on the fermenter and place the sensor in the buffer solutions and calibrate the electrodes with buffers of pH 4 and 7.

f. Fit the pH electrode directly into the vessel port after pH calibration.

g. Ensure that all the tubings which are immersed into the fermentation medium are properly clamped.

h. The vessel is autoclaved for about 20-30 minutes at 121°C using steam passing through boiler.

i. Ensure that the cooling line is connected to the chiller. Chiller helps to decrease the vessel temperature to room temperature.

j. Ensure that all the cables are disconnected.

k. Connect the compressor to the fermenter but all the air lines should have a filter on the line.

Operation

l. Switch on power and set operating temperature, stirring speed and pH.

m. Inject concentrated inoculum from the inoculating port on the top of the fermenting vessel.

n. Draw the sample from port present at the base of fermenter vessel.

Precautions

a. Vessel should not be completely filled with the medium. Ensure that the medium and inoculum volume should not exceed beyond 22L.

b. Exit line should always be opened for the release of gases produce during fermentation.

c. Add inoculum aseptically in the inoculating port to avoid the contamination.

Flow Injection Analysis System *(FIA)*

A sample is injected into a carrier solution which mixes through radial and convection diffusion with a reagent for a period of time (depends on the flow rate and the coil length and diameter) before the sample passes through a detector to waste. A Peristaltic pump/ syringe pump is the most commonly used pump in FIA instruments.

Procedure

a. Turn on the PC and instrument.

b. Select FIAlab for Windows 5.0 and Click on the "FIAlab" icon.

c. Click on Sequential Injection Analysis button.

d. Initiate the instrument by clicking on "Logon" button. The pump will activate and the valve will turn to the waste port (port 1).

e. Next, click on the pump valve to turn the position "In".

f. Set the flow rate and click on "fill" button to aspirate carrier into the syringe.

g. After the pump has finished filling, click on the pump valve again to switch it "Out" in order to access the peripheral ports.

h. Change the flow rate and set the volume after confirming that the multiposition valve is accessing the waste port.

i. Activate the syringe pump to dispense the specified volume by clicking on upper bar.

j. Repeat the above steps (filling the syringe from inlet and dispensing to waste from outlet) until all air has been expelled from the system, leaving carrier in the holding coil of the Lab-on-valve. Also, prime the reagent ports by aspirating 150 μl from reagent #1, reagent #2 and carrier ports (#3, 4 and 6 respectively).

k. Inject 200 μl in the syringe.

l. Data acquisition.

m. Analysis.

Precautions

a. Filter all carrier solvents and samples to remove particulate matter.

b. Accurate pipetting is prerequisite for good results.

Fluorescence Spectrometer

In fluorescence spectroscopy, the excited light/radiation is used to energize the fluorescent molecules by absorbing a photon, from its ground electronic state to one of the various vibrational states in the excited electronic state. Collisions with other molecules cause the excited molecule to lose vibrational energy until it reaches the lowest vibrational state of the excited electronic state. The molecule then drops down to one of the various vibrational levels of the ground electronic state again, emitting a photon in the process that photon is detected by the detector. The emitted light is generally of longer wavelength than the absorbed light.

Procedure

a. Turn on the PC and instrument (Status: Orange light turns green after sometime).

b. Click on the Windows "Start" button.

c. Point to Programs and select Cary Eclipse from the menu.

d. Select the desired application from the second menu (BIO, CONC and SCAN).

e. Select Advanced Reads Application to manually read multiple samples.

f. Start the Concentration application and select Scan to perform calibration.

g. Set up instrument parameters for a new method (Data mode - Fluorescence, Excitation Wavelength, Emission Wavelength, Slit width of the excitation and emission monochromator, Average Time).

h. Set up the display options (Filter and Voltage options).

i. Set up accessories.

j. Set up the calibration (Select the Standards tab, set appropriate units, enter no. of standards being used in calibration and concentration of each in Conc. column, select type of curve fitting required for calibration in Fit Type group, enter correlation coefficient in Min R2 field.

k. Set up the samples (In case of calibration set the No. of Samples field to 0).

l. Set up reporting and printing requirements.

m. Set up storage of collected data.

n. Set up visual system monitoring.

o. Finish set up.

p. Zero the instrument (From Concentration Application window, select the zero button).

q. Read samples or perform the calibration.

r. Save the finished run (*.FMCN file).

Fume Hood Chamber (*GODREJ*)

Fume hood chamber serves to control exposure to toxic, offensive or flammable vapours, gases and aerosols.

Procedure

a. Verify if the fume hood exhaust system and controls are operating properly and providing necessary air flow.

b. Wear appropriate personnel protective apparels such as approved eye protection, lab coat, long pant, safety shoes *etc.*

c. Never place your head inside the fume hood chamber.

d. Prevent the heavy trafficking to fume hood.

e. Clean all spills immediately.

Precaution

a. The fume hood should not be operated with full sash open.

b. Do not place electrical apparatus or other ignition sources inside the hood when flammable liquids or gases are present.

c. Never use for perchloric acid procedures because perchloric acid fumes settle and form explosive crystals.

Gas Cylinders

a. Keep all compressed cylinders in upright position duly secured with a chain to avoid free fall.
b. While operating with compressed cylinders proper care should be taken by taking appropriate spanner for opening and closing the valves. Using wrong tools for this purpose is forbidden.
c. Do not apply grease, wax or any oil *etc.* to compressed gas cylinder valves for lubrication.
d. Use soap water for checking the leakage in the valve of the cylinder/gas distribution system. Note down pressure of cylinders at regular intervals to notice the leaks, if any.
e. Do not drop the cylinders while handling. Use only cylinder trolleys for transferring.

Gas Permeability Tester (*Labthink, Model BTY-B1P, China*)

Gas permeability tester is used to measure permeability co-efficient of the packaging film for O_2 and CO_2 at definite conditions of temperature and humidity.

It is applied to air block and gas permeability test of all kinds of flexible packaging materials such as films, foils, complex films and other plastic filmy materials. It is an indispensable testing instrument in packaging material testing, selection and research work. Keep certain pressure difference between two sides of the specimen under definite conditions of temperature and humidity. Calculate the mass and rate of gas transmission by testing the change of pressure in the lower pressure side.

Procedure

a. The relationship between pressure and time will be displayed on the screen of the computer in real time after test is started and permeation time will be calculated automatically according to permeation ratio.

b. Test will stop when intervened by users or test results are up to the default. The system will decide whether to print test results or operate other data process according to users' requirements.

Precautions

a. Keep the instrument under cool and dust free zone.

b. Ensure the gas cylinders connected properly before start.

c. Check the working condition of valves and regulators of the cylinders.

d. Don't interrupt the instrument when it is working.

e. Switch it off soon after the experiment.

Glass Apparatus

a. Discard irreparable, cracked or broken glass apparatus into properly labeled dustbin. Glass apparatus should be supported and braced in a manner, which will not result in excessive strain.

b. Do not use glass bottles for liquids having vapour pressure above 15 psi at 100°F (1 kg, at 38° C).

c. Discard defective bottles immediately. Wear dry rubber gloves, goggles, gum boots and a rubber apron while handling bottles of acids or other corrosive chemicals. If such chemicals are spilled on the skin or clothing, flood the affected area with copious amount of water.

d. Do not force glass tubes or thermometers in or out of stoppers when dismantaling equipment. If the stopper gets stuck cut them away from the tubes. One should protect his hands by a dust cloth or towel while inserting glass tubes in corks.

e. While using glass apparatus for heating or vacuum/pressure service, check for quality of glass and its withstanding capacity.

Headspace Gas Analyzer/O_2 and CO_2 Analyzer *(902D Dual Track)*

For analysis of oxygen and carbon dioxide in gas flushed (MAP) food packages. It can be used for analyzing oxygen and carbon dioxide in all types of packages.

Procedure

a. Switch it on and press the pump key.

b. Wait till you get the constant readings.

c. Inject the injection probe into the packaging via septum.

d. Press the pump key and wait till you get the constant readings.

e. Note down the readings of $\%O_2$ and $\%CO_2$.

Precautions

a. Protect the probe injection from dust.

b. Don't use the instrument when battery is low.

c. Care must be taken while carrying the instrument as there will be a chance of damage to the injection probe.

Heat Sealer *(Foot Operated)*

Heat sealer is used to seal the packaging material for packaging of fresh and processed food products.

Procedure

a. Green light will glow as soon you switch the heat sealer 'ON'.
b. After 1-2 min you can start sealing the packages.
c. Adjust the timer knob at 2 or 3 pointer mark depending on the packaging material.
d. Keep the package between the heating bars (sealers presses) of the sealer.
e. Press the foot paddle firmly till 2-3 buzzing sound.

Precautions

a. Switch it ON just 1-2 min before use & switch it OFF soon after use.
b. Don't hold the foot paddle beyond 2-3 buzzing sound as it damages the package as well as sealing and heating element.

High Performance Liquid Chromatography *(HPLC)*

HPLC is used for accurate determination of the concentrations of monomeric sugars, oligosaccharides, organic acids, ethanol, furfurals, hydroxymethylfurfurals and other chemical compounds.

Procedure

a. Before starting the instrument, please ensure that the method to be used for determination of required compound is clear, *viz.* flow rate, mobile phase, column temperature, detector, the required temperature and also the column needed for elution of required compounds.

b. Prepare mobile phase and samples in HPLC grade solvents only. Filter the mobile phase and samples through a 0.20 μm filter and sonicate for 15 min.

c. Keep mobile phase in solvent reservoir provided for the purpose.

d. Switch on the main power switch on rear panel of the pump, sampler, column, RI detector and UV detector compartments.

e. The pump runs a series of internal test. Now connect each compartment through computer chromeleon software.

f. Start the chromeleon server monitor by clicking chromeleon icon.

g. Open the purge valve (turn the knob counter clockwise).

i. In chromeleon software open the command dialog box for the pump. Set purge to ON at a flow rate of 3ml/min.

ii. When the specified purge time is elapsed, purge is automatically reset to OFF.

iii. Close the purge valve (turn the valve knob clockwise).

h. Open dialog box for pump, sampler, column, RI and UV detector compartments and connect the system.

i. Gradually increase the flow rate depending upon the type of column used.

j. Switch on the column temperature from column dialog box. Wait until all modules are stabilized. When all the modules have stabilized, green light will appear on all the modules.

Operation

To create a Program and Method

k. Open the program wizard. On the file menu, select new and then select program file.

l. The wizard guides you through program and method creation.

m. After you finish the wizard chromeleon, automatically create a program and method.

To create new Batch

n. Keep the prepared samples in the sampler tray in sampler compartment.

o. Create a sequence using file wizard guide.

p. On the file menu open the Start Batch.

q. Click start to begin the operation.

r. The samples should preferably be diluted with the mobile phase used for separation of compounds. The sample should be filtered through 4mm, 0.45µ syringe filters before the sample is injected into the system. Place the vial in the autosampler tray which is maintained at low temperature. Create a batch and after the parameters are set, start the run and machine will complete the batch itself.

s. After the completion of the batch, turn off the column compartment temperature.

t. When column compartment temperature is in equilibration with the room temperature, start decreasing the flow rate gradually.

u. Disconnect all the compartments.

v. Turn off the main power switch on real panel of the pump, sampler, column, RI and UV compartments.

w. Turn off the system

Precautions

a. Please ensure that the purge valve is open before giving the purge command. The purge valve should be closed, immediately after purging is complete.

b. Always use HPLC grade, filtered and sonicated solvents for mobile phase and sample preparation.

c. The samples and solvents have to be invariably filtered through membrane filters to avoid any contamination in the line or column. The columns are very expensive and extremely sensitive to contamination. Choking of line means that either the line has to be replaced or flushed under high pressure. Such operations can only be performed by trained company professionals.

Hot Air Oven *(MAC Laboratory Equipment & Others)*

It is used for dry heat sterilization and other purposes like moisture estimation.

Procedure

a. Switch on instrument by main.

b. Set the temperature range as desired with the help of digital temperature indicator cum controller.

c. It provides temperature range 50-250°C.

d. After desired temperature is attained allow it to become stable for half an hour.

e. Adjust ventilation ports for regulating ventilation.

f. Always keep the door closed and open it only whenever necessary.

Precaution

a. Never use the instrument without stabilizer.

b. Be careful, while loading and unloading so that sensing element of thermometer does not get damage.

c. If there is a fan, check if it is working properly.

d. Allow to continue heating for an additional 45-60 minutes after the temperature reaches the pre-set degree.

e. Switch off the heat when the time is up.

f. Wait until the temperature falls to 40°C before opening the door.

g. Always follow the instructions of the manufacturer.

Hunter Lab Colorimeter *(MINISCAN XE, 45/0-L)*

The opponent-colour scales give measurements of colour in units of approximate visual uniformity throughout the colour solid. Thus, in the Hunter scale, *L* measures lightness and varies from 100 for perfect white to zero for black, approximately as the eye would evaluate it. The chromaticity dimensions (*a* and *b*) give understandable designations of colour as follows: *a* measures redness when positive, gray when zero, and greenness when negative; *b* measures yellowness when positive, gray when zero, and blueness when negative.

Procedure

a. When measuring samples, it is important to select samples appropriately, use an established measurement method, and handle all samples in a consistent manner.

b. Choose samples that are representative of the material used. If samples are non-representative of the batch or are spoiled, damaged, or irregular, then the result may be biased. When choosing a sample, select randomly and examine the sample to avoid biased results.

c. Present the samples to the instrument in a standard, repeatable manner.

Setup and operation

d. Switch on the handle end of the instrument by pressing the Read (lightning bolt) key.

e. Change the set up if desired to measure in unit other than already set

f. To standardize: first black tile to be read followed by white tile. When standardization is complete, a massage will be displayed indicating that the instrument is Ready to Read.

g. When prompt says Ready to Read Ist sample, place the first sample at the port with the side to be measured towards the port and press the read (lighting bolt) key.

h. Reading displayed.

i. Measure additional samples in the same manner but keep in view that instrument presently is set for four replications at a time, there after it displays average, if required/necessary note down the average. Otherwise simply press the read (lighting bolt) key to script it.

Precautions

a. Please hold the port of MiniScan XE Plus in the centre position.

b. Avoid direct contact of port with the sample.

c. The Petriplate holding the sample should not have empty space.

d. Mimimise the leakage of light to have an accurate reading.

Maintaining MiniScan XE Plus

e. The MiniScan XE Plus requires very little maintenance; just replace or recharge the battery pack as needed, keep the white tile, black glass, and light trap clean, and take reasonable precautions to prevent entry of contaminants into the viewing port.

IR Moisture Analyzer *(Sartorius, Model MA150)*

Procedure

a. Turn on the moisture analyzer.

b. Prepare the sample.

c. Open sample chamber and position a new sample pan.

d. Tare the sample pan: Select the TARE function and confirm.

e. Distribute approx. 2 g of sample on the sample pan. Close sample chamber.

f. Start the drying program.

g. Drying in progress: wait.

h. Read the results.

Safety precautions

a. Use the moisture analyzer only for performing moisture analysis of samples. Any improper use of the analyzer can endanger persons and may result in damage to the analyzer or other material.

b. Do not use this moisture analyzer in a hazardous area/location.

c. Material liable to catch fire or explode may not be used as sample for moisture analysis.

d. The moisture analyzer may be operated only by qualified persons who are familiar with the properties of the sample to be analyzed.

e. Protect the moisture analyzer from contact with liquid.

f. Clean your moisture analyzer in accordance with the cleaning instructions.

g. When setting up the moisture analyzer, leave enough space to prevent heat from building up and to keep your analyzer from overheating:

   i. leave 20 cm around the moisture analyzer

   ii. leave 1 m above the device

h. Do not put any flammable substances on, under or near the moisture analyzer because the area around the heater unit will heat up.

i. Be careful when removing the sample from the chamber since the sample, the heater unit and the sample pan used can still be extremely hot.

j. Prevent excess heat build-up around the analyzer.

Laminar Air Flow (*RESCHOLAR*)

It is used for performing micro biological experiments under aseptic condition.

Procedure

a. Switch on UV light 15-20 min before use.

b. Then switch on the fluorescent light and blower to perform the experiment.

c. Researchers should wear closed front lab coat and gloves.

d. Disinfect work surface with 70% alcohol or other suitable disinfectant before and after use.

e. Minimize air flow disturbances.

Precaution

a. Never switch on the UV light while working on laminar.

b. Turn off the gas knob after completing the work on laminar.

c. Consider the materials necessary for the planned work in the cabinet.

d. Place items into the cabinet so that they can be worked with efficiently without unnecessary disruption of the air flow, working with materials from the clean to the dirty side.

e. Adjust the working height of the stool so that the worker's face is above the front opening.

f. Delay manipulation of materials for approximately one minute after placing the hands/arms inside the cabinet.

g. Minimize the frequency of moving hands in and out of the cabinet.

h. Do not disturb the airflow by covering any of the grillwork with materials.

i. Work at a moderate pace to prevent the air flow disruption that occurs with rapid movements.

Lypholiser

Lypholiser is used for concentration and freeze-drying of liquid samples, such as those of fruit juices or high value products which are degraded by heat treatments. Lypholiser is also needed for lypholisation of microbial cells for long term storage.

Procedure

a. Select the necessary accessories to be used with the basic unit depending upon the product to be lypholized and the purpose for which the lypholization is done, such as ampoule making or product concentration.

b. Connect the unit to the power supply and switch on the unit.

c. Leave the unit running until temperature gauge displays -40°C.

d. Now the unit is ready for use

Operation

e. Mount the freeze drying unit on to the base unit.

f. Close the drain valve in the base unit.

g. Close all the valves of the freeze drying unit (in an upright position).

h. Switch on the main electricity supply to the base unit.

i. Switch on the freezer.

j. When the temperature reaches -40°C, switch on the vaccum pump and attach a pre-frozen flask to the valve by means of suitable flask connector.

k. Open the valve (downward position) slowly to start the drying process.

Switching off the unit

l. When the drying process is complete, do not switch off the vaccum pump but vent each flask individually.

m. After removal of all the flasks, pump can be switch off and the drain/air admittance valve could be opened for releasing the atmospheric pressure.

n. Switch off and defrost the base unit.

Precautions

a. Before switching on the machine, please ensure that all accessories used are dried and clean.

b. Make sure bottom of the condenser chamber is completely dry.

c. Allow the vaccum pump sufficient time to stabilize the pressure in the chamber before connecting flask.

d. When the process is completed do not switch off the vaccum pump, this may result in the flask dropping from the unit.

e. The total loading of the freeze drying unit accessories should not exceed the total capacity of base unit condenser.

f. While using ampoules, always evenly distribute the ampoules on the carrier plate to prevent unbalancing.

g. Also take care to avoid the overfilling of the ampoules so that unnecessary spilling can be eliminated.

Microfuge (*Beckman Coulter Centrifuge 22R*)

Procedure

a. Turn on the instrument.

b. Momentary Operation: Turn rotary switch to "pulse" and rotor will spin only while "start" button is pressed.

c. Timed Operation: Set rotary switch to desired number of minutes and press "start" button. Unit will beep at the end of the timed run. If the cover is lifted during a run, the rotor will stop but the timer will continue to count.

d. Continuous Operation: Set rotary switch to "continuous" position and press "start" button. To stop, set switch back to "pulse" position.

e. Speed Operation: Set rotary switch to required speed from 1000 to 10,000 rpm and press "start" button.

Precautions

a. Use only standard microcentrifuge tubes as described in the instruction manual. Make sure that tubes fit and are positioned properly.

b. Avoid opening cover while rotor is spinning (pilot light is on). If cover is opened, rotor will brake to a complete stop within about 5 seconds.

c. Routinely inspect centrifuge and rotor for cleanliness and damage. Do not run with any signs of damage.

d. Do not use cracked tubes or tubes whose filters or caps do not fit properly.

e. Avoid prolonged exposure to direct sunlight or ultraviolet light. Keep lid closed when not in use.

f. A new rotor should be installed annually.

Microscope

Procedure

a. Ensure that the voltage supply in the laboratory corresponds to that permitted for the microscope; use a voltage protection device, if necessary.

b. Turn on the light source of the microscope.

c. With the light intensity knob, decrease the light while using the low magnification objective.

d. Place a specimen slide on the stage. Make sure that the slide is not placed upside down. Secure the slide to the slide holder of the mechanical stage.

e. Rotate the nose-piece to the 10x objective, and raise the stage to its maximum.

f. Move the stage with the adjustment knobs to bring the desired section of the slide into the field of view

g. Focus the specimen under 10x objective using the coarse focusing knob and lowering the stage. Always turn the stage up towards the objectives while looking from the side and not through the eyepieces, so as to avoid turning it up too far and damaging the objective. Only thereafter do the actual focusing, looking through the eye-pieces, by lowering the stage away from the objectives.

h. Make sure the condenser is almost at its top position. Centre the condenser using condenser centering screws if these are provided in the microscope. For this take out one eye-piece and while looking down the tube, close the iris diaphragm till only pin-hole remains. Check if this is located in the centre of the tube.

i. Open the condenser iris diaphragm to 70-80% to adjust the contrast so that the field is evenly lighted.

Many modern microscopes have pre-centered and fixed condensers. In these no adjustments are required. To reduce glare adjust the opening of the iris diaphragm

j. Adjust the inter-pupillary distance till the right and left images become one.

k. Focus the image with the right eye by looking into the right eye-piece and turning the focusing knob.

l. Focus the image with the left eye by looking into the left eye-piece by turning the diopter ring.

m. Put one drop of immersion oil on the specimen.

n. Change to 100x objective.

o. Increase the light by turning the intensity knob until a bright but comfortable illumination is achieved.

p. Focus the specimen by turning the fine focusing knob.

q. When the reading/observation has been recorded, rotate the objective away from the slide.

r. Release the tension of the slide holder, and remove the slide.

s. Turn off the light.

Precautions

a. If immersion oil was used, wipe it from the objective with lens paper or muslin cloth at the end of each session of use. In general, avoid wiping the objective except when it seems to be dirty. This is sufficient if good quality oil is used (use synthetic oil recommended by the manufacturer).

b. Do not clean lenses frequently. This may cause scratching and chipping of lenses.

c. Always keep the immersion oil bottle capped and free from dust and debris.

d. Use a dropper and not a glass rod to put immersion oil on the slides without touching it.

e. The cover slip should conform to the specifications for the objective of the microscope. Most oil immersion objectives are corrected for cover slip of 0.17 mm thickness.

f. Do not increase the intensity of the light source beyond the maximum permitted value.

g. Do not use bad quality facial tissue or coarse cloth to clean the lens as the coarse fibres can scratch the surface of the lens.

h. Never touch electric bulbs with bare fingers. Natural oil from the skin may burn and darken its surface causing premature decrease in light intensity. Use lens paper to hold the bulb when inserting it.

i. Do not introduce bubbles into the immersion oil by stirring it, or sucking or expelling the oil violently. A bubble under the objective will cause glare and lower contrast, thus reducing the quality of the image.

j. Do not use xylene (or petrol) excessively to clean the lens. Excess oil can be usually wiped off with lens paper or muslin cloth. If good quality immersion oil is used xylene is usually not needed. Avoid using cedar wood oil.

k. Do not exchange objectives of two microscopes unless you are certain that their mechanical tube length specifications are identical.

l. Eye strain should not develop if the microscope is used properly. Never adjust the stage upward while looking through the eye-piece. It will cause the objective to push against the slide and may damage it.

m. Only the 100 objective can be used for viewing under immersion oil. All other lenses are to be used without immersion oil; keep them dry and avoid applying oil or any liquid to these lenses.

Note: In all cases, the manufacturer's manual should be consulted for specific instructions.

Storage

a. Install the microscope on a sturdy, level table. Equipment and instruments which generate vibrations, such as centrifuges and refrigerators, should not be placed on or near this table.

b. The height of the table should be convenient for the use. As and alternative or in addition, and adjustable stool should be made available to make microscopy comfortable. The table should be away from water, sinks, and racks containing chemicals, to prevent damage to the microscope from splashes or spills.

c. Always carry the microscope with one hand supporting the base and the other hand around the arm.

d. Place the microscope in a location from which it need not be moved frequently.

e. Turn the nose-piece to the objective with lowest magnifying power before removing the slide and when the microscope is not in use.

f. If the microscope does not have a built-in light source then the table should be placed near a window away from direct sunlight and arrangements made for the provision of a lamp.

g. In so far as is possible, the microscopy room should be free from dust and should not be damp.

h. If the microscope is to be used every day, do not remove it from the site of installation, provided security is assured.

i. When the microscope is not in use, keep it covered with a polythene or plastic cover and take necessary precautions against fungus.

j. Dust is the worst enemy of the microscope. Always keep the microscope properly covered. Fungus is also a major problem. Always keep the microscope in dry surroundings.

Mini Lab Fermenter (*2.5 L Capacity*)

Mini lab fermenter is used for optimization of different operational parameters such as temperature, pH, aeration level and agitation speed for production of ethanol, enzymes, organic acids *etc*. It can also serve as a feed vessel for the pilot scale fermenter and can be used for scale-up studies.

Procedure

a. Fill the fermenter vessel with required medium by removing the top plate.

b. Fill the reagent bottles with alkali and acid solutions for maintenance of pH.

c. Fit the reagent bottle with a small air filter on one metal plate and fit the silicon tubing to other metal plate.

d. Connect a length of silicon tubing to the peristaltic pump head and clamp the tubing securely from the pump head to the two way inlet.

e. The pH electrode should be connected to base unit. Switch on the fermenter and place the sensor in the buffer solutions and calibrate the electrodes with buffers of pH 4 and 7.

f. Fit the pH electrode directly into the vessel port after pH calibration.

g. Ensure that all the tubings which are immersed into the fermentation medium are properly clamped.

h. Ensure the exit gas line is open and all electrodes have pieces of aluminum foil. Use aluminum foil to loosely cover the drive hub and air filter just to prevent dirt gaining access to the fermenter.

i. Remove the vessel support from the base by unscrewing the heat reflecting shield. Lift the drive bar above the vessel. Ensure all the cables are disconnected.

j. Lift the vessel up and out to release. Place the vessel in autoclave and autoclave it for 121°C for 20-30 min.

k. Ensure enough care when removing the fermenter from the autoclave and allow it to cool properly.

l. Fit the vessel to base unit and replace drive connection. Fit heat reflecting shield and fit the peristaltic pump head to the base unit.

m. Connect all electrodes to their respective cables. Switch on power and set operating temperature, stirring speed and pH on the control panel.

n. Pierce the silicon membrane with syringe needle and inject the required inoculums into the fermentation medium. Close the inoculum port, immediately after inoculation.

o. Remove bottle, cap and replace with new bottle aseptically.

Precautions

a. Ensure that the exit gas line is open and all the electrodes are covered with foil. Use aluminum foil to loosely cover the drive hub and air filter just to prevent dirt gaining access.

b. Ensure maximum care when removing the fermenter from the autoclave.

c. Vessel should not be filled with medium for more than 70% of its volume.

Pipettes

Before using a pipette, be sure that it is the correct size, clean, and inspect the tip for any damages.

a. Place a rubber bulb on the stem of the pipette.

b. An automatic pipette filler may be used.

c. Lower the pipette in to the solution. Allow sufficient depth to fill the pipette above the calibration mark.

d. Apply suction and fill the pipette above the calibration mark.

e. Hold the pipette vertically. The index finger is placed over the upper opening of the pipette and the liquid is allowed to drain to the mark. Excess fluid outside the pipette is drain in to the solution by touching inner wall of the bottle.

f. Transfer the pipette to the receiving container. Drain the pipette freely in a vertical position, and the tip against the side of the container.

g. The pipette must be held in a vertical position when adjusting the liquid level to the calibration line and during delivery. The lowest part of the meniscus, when sighted at eye level should be in level with the calibration line on the pipette.

h. The pipette has been calibrated to deliver its specified volume in a vertical position with a constant rate of delivery.

i. Do not attempt to force the liquid from the pipette at a faster rate than free drainage permits.

j. When using volumetric pipettes the flow of the liquid should be unrestricted and the tip should be touching the inclined surface of the receiving container until 2 seconds after the liquid has ceased to flow.

k. Avoid air bubbles in the pipette.

Quartz Crystal Microbalance (QCM) *(QCM200)*

The QCM measures mass and viscosity in processes occurring at or near surfaces, or within thin films.

This system includes a controller, crystal oscillator electronics, crystal holder, quartz crystals, and software. Anything that has mass can generate a response from a QCM sensor.

Two modes of operation are common for QCM flow cells.

Batch mode

In this mode sample/solvent loaded sequentially and the response of crystal is observed with load/rinse exchanges.

Flow mode

There are two types of flow modes *viz.* an open or closed-flow configuration. Siphon principle is recommended to implement the lowest noise flow setup in open flow mode. The crystal is exposed to a sequence of analytes using Flow Injection Analysis (FIA) and Sequential Injection Analysis (SIA) equipment. Peristaltic pumps are commonly used for closed-flow systems. Pulse-dampening modules are required with these pumps to minimize noise to the frequency signal due to pressure transients and coupled vibrations. In both modes, the flow rate must be kept under 5 mL/min to avoid excessive stress on the sensor and to avoid leaks of liquid over the o-ring.

Procedure

a. Connect the QCM200 Controller to the QCM25 Crystal Oscillator using the Cat-5 cable.

b. Mount a fresh quartz crystal in the Crystal Holder.

c. Connect the Crystal Holder to the 5 MHz Crystal BNC connector of the QCM25 Crystal Oscillator.

d. Connect the RS-232 port to your Host PC loaded with QCM software.

e. Set the Co Compensation switch to Hold and the ten-turn dial to 8.0 (Dry Operation Setting).

f. Turn the Power switch on and check that the Power LEDs in the QCM200 and QCM25 boxes light up. The OSC LED of the QCM controller turns on as soon as proper quartz oscillation is established in the dry crystal. After a short initialization period, an absolute Frequency (F) very close to 5 MHz (nominal dry frequency) should be present on the front panel Measurement display.

g. The QCM200 System is now ready for mass and viscosity measurements.

h. For operation under liquid immersion, record the "dry" frequency readings and immerse the entire crystal in the solution of interest.

Precautions

a. When using a flow cell, avoid air bubble trapping in the measurement volume while filling the flow chamber for the first time.

Refrigerators

Precautiuons

a. Refrigerator must be placed so that sufficient air can pass the condenser (at the back of the refrigerator) for exchange of heat and also to facilitate cleaning of the condenser.

b. The refrigerator door must seal perfectly to prevent warm outside air from entering the cool chamber.

c. The refrigerator must have good insulating walls.

d. Do not overload.

e. Remove all sample that are not worth.

f. Do not put any sample without proper labelling.

Daily Checks

a. Check temperature daily. It should not exceed 12°C. Application of battery driven mobile or stationery thermometers are recommended, preferably those including continuous printing or plotting of temperature measured when heat-sensitive reagents are stored for long periods.

Monthly Checks

a. Clean cool chamber and defrost the evaporator monthly.

b. Clean refrigerator from the outside.

c. Clean condenser for dust.

Refrigerated Centrifuge *(Model MP 400R)*

It is used to separate liquid solutions (at refrigerated temperature with different speed and time) into layers based on their mass. More massive components of the liquid are pressed toward the outside of the centrifuge with more force, so they settle to the bottom of the test tube. Less massive components settle in layers higher up, with the least massive at the top.

Procedure

a. Place the samples in the slots of rotor of appropriate size and close the lid.

b. Switch the machine on.

c. Select the program by pressing the program key and set the RPM, temperature, time for the experiment by using arrow keys.

d. Press the start key (press the star key '*' if you require cold centrifuge).

e. Wait till centrifuge completely stops before opening the lid.

Precautions

a.. Don't open the lid till it stops completely.

b. Clean the machine after use.

c. Use appropriate tubes with caps as per the slot size of the rotor.

d. Close the lids properly.

e. Don't run the machine at high temperature (>15°C).

Sausage Filler

These machines are used for filling all types of meat batter in containers such as casings, glass jars, cans *etc.* The most common type of filling machine in small and medium size operations is the piston type. A piston is moved inside a cylinder forcing the meat material through the filling nozzle (funnel, stuffing horn) into the containers. Piston stuffers are either attached to the filling table (manual) or designed as floor models (hydraulic).

Procedure

a. Open the screws of sausage filler.

b. Keep the material in the cylinder.

c. Close the sausage filler.

d. Attach suitable funnel and casings.

e. Switch on the main.

f. Push down the foot paddle for filling.

g. Stop the foot paddle after completion of filling.

h. Put off the main.

i. Open the screws of sausage filler.

j. Take out the remaining material from cylinder.

k. Use dry paper or cloth to remove all the meat and ingredients attached to the sausage filler.

l. Soak a sponge duster in liquid detergent and use it to clean the sausage filler.

m. Clean the filler using a wet cloth.

n. Finally use wet cloth dipped in hot water for final cleaning.

o. Dry the sausage filler with a clean dry cloth.

Sirman Meat Grinder

A meat grinder is a machine used to force meat or meat trimmings by means of a feeding worm (auger) under pressure through a horizontally mounted cylinder (barrel). At the end of the barrel there is a cutting system consisting of star-shaped knives rotating with the feeding worm and stationary perforated discs (grinding plates). The perforations of the grinding plates normally range from 1 to 13mm. The meat is compressed by the rotating feeding auger, pushed through the cutting system and extrudes through the holes in the grinding plates after being cut by the revolving star knives. Simple equipment has only one star knife and grinder plate, but normally a series of plates and rotary knives is used. The degree of mincing is determined by the size of the holes in the last grinding plate. If frozen meat and meat rich in connective tissue is to be minced to small particles, it should be minced first through a coarse disc followed by a second operation to the desired size.

Procedure

a. Please use semi frozen meat for grinding.
b. Wash all parts of grinder in warm water (70°C) and dry them.
c. Assemble all parts of the grinder carefully.
d. The position of cutting blade should be proper.
e. Place the sieve plate over the blade.
f. Screw cap the plate. Don't over tighten.
g. Place Hopper plate on head and fix into position.
h. Switch on the main power plug.
i. Start grinding the meat.
j. After use Switch off the main.
k. Disassemble all parts.
l. Wash the grinder and its parts with tap water and then use liquid detergent solution for cleansing. Final washing should be with warm water.
m. Dry all parts and assemble again.

Sonicator *(Sonics Vibra Cell)*

Sonication involves dissipation of sound or ultra sound energy to homogenize or agitate particles in a sample for various purposes. In biological applications, sonication may be sufficient to disrupt or deactivate a biological material.

Modes of Operation

Timed Operation

Turn timer control to the desired processing time. Set Duty cycle control to constant and output control to 1. With the tip of the horn immersed about ¼ to ½ inch in the solution to be processed, turn power switch ON. To increase ultrasonic intensity turn output control to the next highest number, pausing at each setting in order to evaluate results until the optimum setting is reached.

Continuous Operation

Turn Timer control fully to the hold position. Set Duty cycle control to constant and power switch ON. Adjust output control to the desired required intensity.

Pulsed operation

Use of the Duty Cycle control enables the ultrasonic energy to be pulsed, one pulse per second. The control settings from 10% to 90% of each change the duration of the pulse from 10% to 90% of each second respectively.

Pulsing retards the rate of temperature increase that results from the ultrasonic activity. To operate in the pulsed mode, set the Timer Control to Hold if the process is not to be timed or if a processing time longer than 15 mins. is required. Set Duty Cycle Control to the required pulse, with a lower setting giving a shorter pulse, resulting in a slower temperature rise in the solution. Switch power to 'On' and adjust Output Control to the required intensity.

Procedure

a. Pre-check sonicator equipment.

b. Set the parameters of amplitude, time, temperature, energy level and pulsing based on probe used and sample characteristics and save as protocol.

c. Work on your samples as per your protocol.

d. Clean the probes and converters after use, first with 10N NaOH, then 70% Ethanol, and last with double distilled water. Clean with acetone if fat is there in sample.

Precautions

a. Users must be trained on sonicator first.

b. Wear ear protection, gloves, lab coat and glasses.

c. Use in the fume hood because of aerosols.

d. When spill occurs, follow spill protocol according to specific samples.

e. If emergency occurs, switch off the sonicator directly or unplug.

Spectrophotometer *(Dual Beam UV- VIS)*

The dual beam UV-VIS spectrophotometer is used for determination of sugars, carotenoids, pigments, anti-oxidants, enzyme assays *etc.* It is a basic instrument used for analysis of different food constituents. A spectrophotometer consists of two instruments, namely a *spectrometer* for producing light of any selected color (wavelength), and a *photometer* for measuring the intensity of light. The instruments are arranged so that liquid in a cuvette can be placed between the spectrometer beam and the photometer. The amount of light passing through the tube is measured by the photometer. The photometer delivers a voltage signal to a display device, normally a galvanometer. The signal changes as the amount of light absorbed by the liquid changes.

If development of color is linked to the concentration of a substance in solution then that concentration can be measured by determining the extent of absorption of light at the appropriate wavelength. For example hemoglobin appears red because the hemoglobin absorbs blue and green light rays much more effectively than red. The degree of absorbance of blue or green light is proportional to the concentration of hemoglobin.

When monochromatic light (light of a specific wavelength) passes through a solution there is usually a quantitative relationship (Beer's law) between the solute concentration and the intensity of the transmitted light, that is, $I=I_0 10^{-kcl}$

where I_0 is the intensity of transmitted light using the pure solvent, I is the intensity of the transmitted light when the colored compound is added, c is concentration of the colored compound, l is the distance the light passes through the solution, and k is a constant. If the light path l is a constant, as is the case with a spectrophotometer, Beer's law may be written, $I/I_0=10^{-kc}=T$

where k is a new constant and T is the transmittance of the solution. There is a logarithmic relationship between transmittance and the concentration of the colored compound. Thus, $-\log T=\log 1/T=kc=$ ***optical density***

The O.D. is directly proportional to the concentration of the colored compound. Most spectrophotometers have a scale that reads both in O.D. (absorbance) units, which is a logarithmic scale, and in % transmittance, which is an arithmetic scale. As suggested by the above relationships, the absorbance scale is the most useful for colorimetric assays.

Procedure

Initialization

a. Turn on the instrument and it begins to initialize, and it takes three to five minutes to initialize.

Gain access to the main menu after initializing

b. Gain access to the photometric mode: Press [ENTER] to gain access to the photometric mode.

c. Gain access to measurement menu: Press [START] to gain access to measurement menu.

d. Set wavelength: Press [GOTOλ], then input the working wavelength. For example, if we need to measure at 460 nm, input 460, and press [ENTER] to confirm.

e. Gain access to parameters setting: Press [SET] to gain access to the parameters setting page, then, press [▼] and move the cursor to the "Cell Setting". Then, press [ENTER] key to confirm.

f. Set Cell Number: Press [▼] to move the cursor to the "Cell Number", Then press [ENTER] to choose the cell number. (It depends on the cells you use. If you use 2 cells, you should choose 2).

g. Measurement: Press [RETURN] key twice to return to the photometric menu. Put the blank into the cell No. 1, and put the sample into No. 2 cell, then press [ZERO], and then press [START] to begin measurement.

h. When needed to measure the next sample, just put the sample into the sample compartment, and press [START]. If you need to change the wavelength, press [GOTOλ] to input the required wavelength. Note: Press [ZERO] to perform correction after you change the wavelength. If you use the sample cell number to measure each time, you ignore the step 5 and step 6.

i. Finish measurement: After measurement, press [PRINT] to print the results, if needed. Please write down the results if the printer is not available. The results will be lost if we quit and turn off the instrument. Make sure to take out the cell from the cell compartment. Press [RETURN] to return to the instrument main menu and then turn off the power.

Precautions

a. Check if there is any spillage on the floor of the sample compartment before and after use. Wipe off any spilled liquid immediately to prevent corrosion of the compartment, which may cause erroneous result and damage the instrument.

b. There are four dust filters at the bottom of the instrument. Generally speaking, the user should clean them every three months. But if in the places under severe conditions such as with much sand, it should be cleaned once a month. Use tap water to clean the filter.

c. Do not spill solution on the cover. Otherwise, the cover will be blemished. If the solution spills on the cover, please clean it with wet towel. Avoid using organic solution to wipe the cover.

Spectrometer NIR Analyzer AOTF *(5030 Miniature; Range: 1200-2200nm)*

Procedure

a. This spectrometer is based on the SNAP32 software consists of several application programs that use standard windows graphical user interface (GUI) to simplify data acquisition and processing routines. It is used to acquire, process, view and edit spectral data.

b. The acquire module of SNAP32 system is a spectra acquisition module. Upon starting the acquire program, the system prompts a user to enter his/her password to access to the module according to the security level assigned by the administrator. After successful login, the application window will appear containing the buttons *viz.* start acquisition, set parameters.

c. Click on the set parameters buttons. This will initiates a window having different options for the wavelength range, wavelength increment, scan average, acquisition mode *etc.*

d. Select the automatic mode for acquisition; locate the file location to be saved. Give the output file name according to your sample and also add the file description and press ok.

e. After setting the acquisition parameters, click the button "Start acquisition". This will start the scanning of the sample.

f. The output graph of transmittance versus selected wavelength range will be displayed.

g. After spectral acquisition, press the exit button.

h. The acquired data is then saved and transferred to the excel worksheet or directly to the chemometric software for further processing.

Precautions

a. Never touch the lens of the spectrometer.

b. Always use soft cloth to clean.

c. Keep the spectrometer covered while not in use to avoid dust accumulation on the lens surface.

d. Always keep the batteries fully charged to avoid any hurdle during acquisition.

Spectrometer NIR USB *(EPP2000 NIR InGaAs512; Range: 900-1800nm)*

Procedure

a. Use the Setup → Unit calibration coefficients menu to enter the 3 coefficient values found on the spectrometer label.

b. Select unit 1 with physical address 1.

c. Taking a Dark scan for the Scope mode becomes important with detector integration times well above 250ms (above 1/4 second).

d. Always take a "new dark" after changing system parameters such as samples to averaging, smoothing, and detector integration.

e. To take a dark scan, you must block the light signal input to the spectrometer or turn off the light for a moment.

f. Before "viewing" Absorbance or Transmittance, first setup some basic system parameters and save a "dark" and "reference" scan. This is performed in the default "Scope" view mode.

g. Setup and turn on reference light source so that you are now viewing a bell shape curve in Scope mode.

h. For reflectance, using a fiber probe, hold at 45° to your white standard about 1/4 inch away. For cuvette holders, use cuvette in place with no sample. For dip probes, use your solvent solution for reference.

i. For reflectance, just move the tip back from the white standard (reference) surface.

j. It recommend that start with (and setup) the following configuration until you are familiar with the options.

Use the Setup menu

i. Detector integration rate = 50 millisec

ii. Number of Scans to average = 5 (if below 100 ms)

iii. Pixel resolution (smoothing) = 0 (NONE)

iv. Temperature Compensation = 1 (ON)

Note: if the first 3 items are changed, then you MUST AGAIN save a new "dark" and "reference" scan.

k. Adjust the detector integration until the bell curve fills 90 percent of the graph. The longer that the detector integrates, the larger the input signal.

l. Perform a File → Save → Reference from the menus or click the yellow light bulb icon on the toolbar.

m. Now block the light at light source. If can't then will have to turn it off for a moment.

n. Perform a File→Save→Dark or click the dark light bulb on the toolbar.

o. At this point ready to view Absorbance or Transmission using the View menu selection.

p. Absorbance and Transmission should appear as flat lines.

q. Insert your sample and observe the response in real-time.

r. Now save data to disk or print the sample graph.

Precautions

a. It is better to take help of a scientist who knows the operation of NIR spectrometers.

b. Keep the temperature of the area of the measurements and samples as constant as possible.

c. Choose correct sample holders depending upon the nature and amount of samples.

Spectrometer VIS NIR USB

(Avantes, AvaSpec-2048-USB2; Wavelength range: 300-1100nm; Resolution: 0.04-20nm, depending on configuration)

Procedure

a. Connect a fiber or probe to the light source and to the spectrometer input port/s and set up the experiment for taking a reference spectrum. Follow the instruction of operation manual of this instrument.

b. Start the AvaSoft software and click the green start button to start measurement of a spectrum.

c. Present optimal smoothing and store on board in the EEPROM.

d. Now turn on the light source.

Note: Usually some sort of spectrum may be seen on the screen, but it is possible that too much or too little light reaches the spectrometer at the present data collection settings. Too much light means that, over a certain wavelength range, the signal is saturated shown as a straight line at the maximum counts and the appearance of the label "saturated" in the status bar of the spectrometer channel. This can usually be solved by a shorter integration time. The integration time can be changed in the main window, in the white box below the start/stop button or the Auto-configure Integration time button can be used. The new integration time can be entered in the box, and will take effect once you press the 'Enter' key or leave the box. Try to adjust the integration time, such that the maximum count over the wavelength range is around 90% of the full ADC scale, which is 14750 counts for the 14bit ADC, 59000 counts for the 16bit ADC). When at minimum integration the signal is still too high, an attenuator, a neutral density filter or fibers with a smaller diameter may be used. When not enough light reaches the spectrometer, likewise a longer integration time can be entered.

e. When a good spectrum is displayed, turn off the light source.

f. Now save the Dark data. This can be done by File-Save-Dark from the menu or by clicking the black square on the left top of the screen with the mouse.

Note: Always use Save Dark after the integration time has been changed.

g. Turn on the light source again. Save the present spectrum as a reference by choosing File Save-Reference from the menu or by clicking at the white square (next to the black one).

Note: Always use Save Reference after the integration time has been changed.

h. Now the Transmittance/reflectance (T button) or Absorbance (A button) spectra can be obtained online. To have a better look at the amplitude versus wavelength, click the cursor button. A vertical line is displayed in the graph. If the mouse cursor is placed nearby this line, the shape of the mouse cursor changes from an arrow to a 'drag' shape. If this shape is displayed, the left mouse button can be used to drag (keep left mouse button down) the line with the mouse towards a new position. Moving this line shows the corresponding values of wavelength and amplitude in the main screen.

i. By clicking the red stop button, the data acquisition stopped and the last acquired spectrum is shown in static mode.

Note: The data acquisition can be started again by clicking the same button, which now shows a green 'Start'.

j. To save the spectrum (in the mode chosen before), choose File-Save- Experiment from the menu, or click the Save Experiment button from the button bar.

k. To improve the Signal/Noise ratio, take average of number of spectra. To do this, the value in the white average box in the main window (next to integration time) can be increased. The new value will take effect when you press the 'Enter' key or leave the box.

Functions of buttons

Save reference and dark buttons

The reference button is the white button at the left top of the screen. It needs to be clicked to save the reference data. The same result can be achieved with the option File-Save Reference. The dark button is the black button at the left top of the screen. It needs to be clicked to save the dark data. The same result can be achieved with the option File-Save Dark

Save experiment button

By clicking the Save Experiment button an experiment is saved. The same result can be achieved with the option File-Save Experiment.

Scope button

By clicking the S button, the data will be presented in Scope Mode. The same result can be achieved with the option View-Scope Mode.

Absorbance button

By clicking the A button, the data will be presented in Absorbance Mode. The same result can be achieved with the option View-Absorbance Mode.

Transmittance button

By clicking the T button, the data will be presented in Transmittance Mode. The same result can be achieved with the option View-Transmittance Mode.

Irradiance button

By clicking the I button, the data will be presented in Irradiance Mode. The same result can be achieved with the option View-Irradiance Mode.

Auto Configure Integration time button

After this button is clicked, AvaSoft starts searching for an optimal integration time. Depending on the maximum counts in the last scan, the integration time will be increased/decreased automatically until a scope signal of about 90% of the full scale is measured. During the search routine, the changes in integration time can be followed in the integration time edit box in the edit bar. A dialog will display the new integration time if the search is ready. The routine can be aborted by clicking the button again before the search has ended.

Precautions

a. Use this instrument in a clean laboratory environment.

b. The unit is designed for operation in dry rooms only.

c. The unit should be situated so that its location or position does not interfere with its proper ventilation.

d. The unit should be situated away from radiators, hot bodies, ovens or other heat sources.

Spectrophotometer UV-VIS (*Model Spectroscan 80DV*)

It is used for the analysis of foods and agricultural products for quantification of chlorophyll content, extent of browning *etc*. by measuring optical density of the solution at different wavelengths.

Procedure

a. Switch on the instrument and after all check-up choose option no.1 *i.e.* photometric evaluation from the 5 options *viz.*, photometric measure, spectrum measure, quantification, DNA/protein, utilities.

b. Set the parameters for the test: a) press F1 for setting wavelength with the help of button 2 then press return key, b) then press F3 and set the parameter as:

   i. sample module/total cells=8

   ii. cells used/drive no.= [blank + number of samples]

   iii. cell/blank=yes [press 3 to change from 'No' to 'Yes']

   iv. move cell=1

   v. reset followed by pressing the return key

c. After setting the parameter for the test, set absorbance to zero by pressing the key A/Z.

d. When absorbance is zero, press the start key and not down the first reading only.

e. Always place the blank first and then samples.

Precautions

a. Switch the spectrophotometer 'ON' at least half an hour before use.

b. Rinse the cuvettes properly before and after use.

Surface Plasmon Resonance (SPR)

Surface plasmon resonance (SPR) biosensor has become a standard technology for measuring kinetics of bimolecular interactions without the need for labeling and gives the real time data. During the SPR measurement, one partner of the interaction to be studied (called a ligand) is immobilized covalently to the sensor surface and the other partner is passed over the chip in solution (called an analyte). The operating principle of SPR is based on the interaction between electromagnetic wave of incident light and free electrons in a conducting surface layer. It measure the refractive index changes near the sensor chip surface produce by binding events involving a target analyte.

Procedure

a. Switch on the instrument and wait for 5 min. for the instrument to stabilize.

b. Start the computer.

c. Click on the icon ProteOn manager to initiate the software.

d. Wait till the computer establishes communication with the instrument.

e. Wait for the main menu screen.

f. Select the appropriate option from the menu to start the experiment.

g. Insert the corresponding chip in the instrument.

h. Before starting the experiment, select the appropriate washing procedure for the chip and follow the subsequent instructions.

i. After washing of the chip, load the vials of antigens, antibodies, buffers *etc.* in the sample tray.

j. Select the required experiment from the list of saved experiments or build a new experiment.

k. Load the experiment.

l. Run experiment.

m. After the experiment is complete give 'post run washing' and exit the software.

n. Wait for the software to shut down.

o. Switch off the instrument.

Precautions

a. Spin all solutions to be placed in autosampler compartment to degas.

b. Accurate pipetting is prerequisite for good results.

Thickness tester (*Model CHY-C2, China; Accuracy: up to 0.1μ*)

Thickness tester is applicable for precise thickness test for plastic film, foil and other similar materials.

This tester employs the most advanced technology in the world, to guarantee the accuracy of the test result and high consistency of several test results. Easy operation is approximately automatic, decreasing the influence to test result caused by manual factors. Print out test result of single test. Statistic, analyzes, and prints out test result of batch of tests. Contacting area, test pressure, and moving speed confirm to relative standards strictly.

Procedure

a. Calibrate the instrument using standard specimen.

b. Press reset key then enter key to display menu.

c. Select and set the required conditions from the menu *viz.*, number of test readings, manual/ automatic *etc.*

d. Press the enter key and note down the readings of maximum, minimum, and average thickness in 'mμ'.

Precautions

a. Keep the instrument under dust free zone by covering the instrument.
b. Don't interrupt the instrument when it is working automatically and wait till it stops itself.
c. Don't use hard material for measurement.
d. Abstain from touching the instrument probe while it is moving against base plate.
e. Clean the instrument after use.

Ultra Low Temperature Freezer

Ultra low temperature freezer is used for long-term storage of microbial cells, live cultures, expensive enzyme substrates, membranes and any other sample which requires storage in the range of -50°C to -86°C.

Procedure

a. Connect the unit to the power supply using power cord provided.
b. Turning on: turn the key provided (one quarter turn to the right) on lockable panel.

c. Set the on/off circuit breaker to on position.

d. The temperature display illuminates immediately.

e. Temperature setting: Press the set temperature key, display indicate zero. Set the temperature using numerical key. If you make any input error use C key to clear the display.

f. When the correct temperature is displayed, press the E key to enter the data. The SET TEMP lamp will go off.

g. To activate the alarm, put the battery switch to the on position and test its operation by pressing alarm test/ mute key on the display.

Note:

h. The freezer compressor will at most operate for approximate 1 min after connection of the power supply. Temperature and alarm setting can be adjusted immediately.

i. When the equipment is first switched on it will take 6 hrs to attain the temperature of -86°C.

j. If there is problem in opening the door, press the manual plunger on the outside of the air vent in order to break the ice formation.

k. Defrosting is done by unplugging the freezer from electrical supply.

Precautions

a. After an extended period of operation, defrosting is important.

b. Do not attempt to scrape the ice with a sharp instrument.

c. Do not use abrasive cleaner or solvent to clean.

Vacuum Oven

Procedure

a. Turn the pump on (the switch is located on the front on the pump).

b. Push the 'main power switch' and set the temperature of the oven using the 'manual temperature control'. Open the vacuum valve and wait until the temperature of the oven stabilised.

c. Put the sample into the oven. Closed the 'vacuum valve', open the N_2 shut-off valve and the 'N_2 'needle valve'. Vent the oven. Open the door, insert samples, close the 'N_2 'needle valve' and close the door.

d. Open the vacuum valve and wait for the pressure to drop.

e. Take your sample out. Close the vacuum valve, open the door and remove samples. Close the door.

f. Close the N_2 shut-off valve, set the manual temperature control to 'zero', and turn off the power.

Precautions

a. Apply vacuum greases to door gasket frequently.
b. Keep side and back vents clean of dust for maximum air circulation.
c. Clean up spills in oven as soon possible after they occur to avoid having the material being backed on. Use a soft cloth and hot detergent water when the oven is cool.

Vacuum Packaging Machine *(Gas Flushing Type)*

It is used for "Modified Atmosphere Packaging" by two ways. In one way, it is used for vacuum packaging by removing atmosphere entirely from the package containing food products. It also used to alter the atmosphere by gas flushing (generally CO_2 and N_2) referred to as gas flushed packaging. In each case, the objective is to extend the shelf life of perishable foods, while at the same time maintain or improve the quality of the product.

Procedure

a. Check the gas cylinder connection.

b. Switch the machine 'ON'.

c. Set the vacuum time (35-40 sec), heating time (1.5-2.0 sec), inflating time (1.5-2.0 sec for gas filling & 0.0-0.1 sec for no gas filling), and temperature (low-medium) by using up & down arrow keys.

d. Keep the packaging on both side of the heating material in the heating chamber.

e. Start the gas to flow just before the process starts and regulate the gas flow.

f. Press the cover of the heating chamber till the process completes.

Precautions

a. Use the gas just before the process starts, regulate the gas flow and stop the gas flow as soon the process completes.

b. Make sure that proper connection of cylinder and working condition of regulator before use.

Water Bath

It is used for providing preset temperature condition and temperature equilibration in experimental protocols.

Procedure

a. Ensure the minimum water level to fully immerse the heating element to avoid its burning.

b. Switch on the instrument by main.

c. Set the desired temperature range with help of controlled digital temperature indicator cum controller.

d. After desired temperature is attained allow it to become stable for half an hour.

e. Use distilled water to prevent corrosion of walls.

f. Change water frequently.

Precautions

a. Water bath should never be switched on without water.

b. Always use distilled water.

c. The level of water in the water bath should be above the level of the solution in tubes to be incubated.

d. Temperature should be maintained daily. Prior to incubating the tubes, check the temperature at the centre and four corners of the water bath.

e. When incubating open containers, vials or tubes the water bath lid should be removed to prevent contamination and dilution of the incubated material by condensed water.

f. The water bath must be refilled to prevent growth of algae and bacteria.

g. Preventive maintenance: The bath should be filled with distilled water. The unit can be cleaned with commercial stainless steel cleaning agents. The heating element should always be kept clean.

h. Objects liable to rust must not be placed in the bath.

Water Vapour Permeability Tester/WVP Tester

(Model TSY-W3 Electrolytic Detection Method Water Vapour Permeability Tester)

TSY-W3 Electrolytic Detection Method Water Vapour Permeability Tester is applicable for water vapour permeability test of various packaging materials, including plastic films, laminated films, back sheet, high barrier materials, sheets, foils and waterproof membrane. The water vapour permeability rate testing can control and adjust the technical indexes of materials, so as to meet application demands.

Procedure

a. Place the specimen between the constant temperature dry and humid chambers.

b. Owing to the pressure differences, saturated water vapour will permeate through specimen from the high humidity chamber to the low humidity chamber and is then carried to the sensor, which would generate electrical signal at the same time.

c. Through electronic signal analysis and calculation of the sensor, water vapour permeability rate and water vapour permeability coefficient can be obtained.

Technical Indexes

i. Test Range: 0.001~50 g/m^2.24h (Routine) & 0.01~1,000 g/m^2.24h (Optional).

ii. Temperature Control Range: 5°C~95°C.

iii. Temperature Accuracy: ±0.1°C.

iv. Humidity Range: 0%RH, 2%RH~98.5%RH, 100%RH (Standard 90%RH).

v. Test Area: 38.48 cm^2.

vi. Specimen Thickness: <1mm (Thicker specimen needs accessories).

vii. Specimen Size: ϕ100 mm.

viii. Carrier Gas: 99.999% high pure nitrogen.

ix. Carrier Gas Flux: 100 ml/min.

x. Power: AC 220V; 50Hz.

15

Safe Storage of Laboratory Equipment

15.1 Store glass tubing and rods in a horizontal position with no place protruding over the coping.

15.2 Store delicate glassware in separate cartons clearly marked for ready identification.

15.3 Store glassware not higher than a person can reach easily without standing on a stepladder.

15.4 Store heavy pieces on lower shelves. Store tall pieces at the back, smaller ones towards the front of the shelf.

15.5 Special metal or plastic containers should be used to transport bottles of acids, alkalies etc.

16

Common Chemical Solution/ Reagents

The following is a list of the solutions and chemicals generally required in biological laboratory. For many solutions, directions are given for a molar solution and the user is left to dilute to the appropriate concentration for their needs.

Acetic acid (MW 60.05)

Glacial acetic acid is 99.6% (w/v) acetic acid, and is 17.4 M.

1 M

Add 57.5 ml of glacial acetic acid to 800 ml of water and then make to 1 litre with water.

0.05 N

Add 2.87 ml of glacial acetic acid to 800 ml of water and then make to 1 litre with water.

0.9 M

Add 54 ml of glacial acetic acid to 800 ml of water and then make to 1 litre with water.

7% (w/v)

Add 70 ml of glacial acetic acid to 800 ml of water and then make to 1 litre with water.

45% (w/v)

Add 450 ml of glacial acetic acid to 500 ml of water and then make to 1 litre with water.

Acetic Acid/Butanol/Water (15:60:25)

Combine 150 ml of glacial acetic acid, 600 ml of n-butanol and 250 ml of water.

Aceto-orcein

Add 2.0 grams of orcein to 45 ml of glacial acetic acid. Bring to a boil and continue to heat until completely dissolved. Cool and add 55 ml of distilled water. Filter prior to use. Note: Some early investigators added a of an iron salt (such as ferric citrate) as a mordant. It tends to increase the intensity of the aceto-orcein stain. The same reaction can be had by chopping plant material with an older steel (not stainless) razor blade.

Acid alcohol

Add 1.0 ml of concentrated HCl to 100 ml of 70% (v/v) ethyl alcohol.

Acid Orcinol Reagent (0.1% $FeCl_3$ in 10% HCl)

Add 0.1 grams of $FeCl_3$ to 50 ml of 10% (v/v) HCl and make to 100 ml with 10% HCl.

Acrylamide Solutions

Acrylamide solutions for PAGE are given as total concentration of acrylamide (acrylamide + bisacrylamide) and the amount of cross linker (bisacrylamide). This is listed as the T:C ratio. For example, a 10% gel (10%T:5%C) would contain a total of 10 g rams of acrylamide per 100 ml, and would be composed of 5 grams of acrylamide and 5 grams of bisacrylamide. Usually, a stock solution of 30% acrylamide is produced containing 0.8% bis-acrylamide. Many investigators use 30 grams of acrylamide plus 0.8 grams of bis-acrylamide per 100 ml of water, but 29.2 grams of acrylamide plus 0.8 grams of bis would be technically correct. In practice, it makes little difference since the gels are diluted to 10% or less. The 30% stock solution is filtered through a 0.45 μ filter and stored at 4°C in the dark. For use, the stock solution is diluted with an appropriate buffer (usually a 2X Tris-HCl). The stock solution is stable for about one month. Discard after this period.

Acrylamides in their monomeric form are neurotoxic. Polymerize all acrylamide solutions prior to disposal.

SDS,-mercaptoethanol and a tracker dye (bromophenol blue) are added at various points.

Alcian blue (MW 1300)

0.001 M

Dissolve 0.13 grams of Alcian Blue 8GX in 100 ml of water.

Alcohol Orcinol Reagent (10% orcinol in 95% ethanol)

Dissolve 1.0 gram of orcinol in 95% ethanol to a final volume of 10 ml.

Alkaline Distilled Water

Add one pellet of NaOH to 1 litre of distilled water.

Alkaline Solution for G-banding

Dissolve 2.8 grams of NaOH and 6.2 grams of NaCl to a final volume of 1 litre with water.

p-Aminosalicic Acid (PAS MW 175.1)

6% (w/v)

Dissolve 6.0 grams of PAS to a final volume of 100 ml with water or buffer.

Ammonium acetate (MW 77.08)

0.1 M

Add 7.708 grams of Ammonium acetate to a final volume of 1 litre of water.

Ammonium persulfate (MW 228.2)

10% (w/v)

Dissolve 1.0 grams of ammonium persulphate to a final volume of 10 ml with water. Mix fresh, prior to use as a catalyst for PAGE. Normally, about 50 µl of ammonium persulfate is added to each 15 of gel solution for polymerization.Dissolve 0.13 grams o f Alcian Blue 8GX in 100 ml of water.

Ammonium sulphate (MW 132.14)

2% (w/v)

Add 2 grams of ammonium sulphate to a final volume of 100 ml water.

4.1M (sat.) 0.001 M

Dissolve 542 grams of ammonium sulphate to a final volume of 1 litre.

n-Amyl alcohol (Pentanol $C_5H_{11}OH$ MW 88.15)

Density = 0.8144 grams/ml 0.38 M

The amyl alcohol can be weighed (33.5 grams) or measured volumetrically by using the density. That is, 33.5 grams 0.8144 grams/ml or 41.1 ml of n-amyl alcohol. Weigh or measure the appropriate amount and dilute to a final volume of 1 litre with water.

Amylase, buffered pH 7.0

1% (w/v)

Dissolve 0.5 grams of amylase to a final volume of 50 ml with 0.01 M sodium phosphate buffer, pH 7.0.

Ascorbic acid (MW 176.12)

2 mM

Dissolve 35.2 mg of ascorbic acid to a final volume of 100 ml with water.

ATP (Adenosine triphosphate, MW 507.21)

5 mM

Dissolve 254 mg of ATP to a final volume of 100 ml with water or buffer.

Baker's Formalin

Add 1.0 gram of calcium chloride, 1.0 gram of cadmium chloride and 10 ml of concentrated formalin to 75 ml of water. Make to a final volume of 100 ml with water.

Benzoic acid (MW 122.12)

8 mM

Dissolve 98 mg of benzoic acid to a final volume of 100 ml with water or buffer.

Biuret Reagent

Add 1.50 grams of $CuSO_4 \cdot H_2O$ and 6.0 grams of sodium potassium tartrate to 500 ml of water. Separately make 300 ml of 10% (w/v) NaOH by dissolving 30 grams of NaOH to a final volume of 300 ml with water. Combine the two solutions in a 1 litre volumetric, swirl to mix and make up to 1 litre with water. Store the final solution in a dark, plastic bottle. Discard if black or red precipitate forms.

Bovine Serum Albumin (BSA)

There are many grades of BSA available and care should be taken when using this protein. For routine protein concentration standards, a 96-99% pure fraction may be used. For tissue culture, RIA, or molecular weight standardization, BSA should be obtained which is extracted and purified specifically for that purpose.

1% (w/v)

Dissolve 0.5 grams of BSA to a final volume of 50 ml in water or buffer.

Bradford Protein Assay

This procedure uses an absorbance shift in an acidic Coomasie Blue solution. It contains methanol and solubilising agents and is very reliable.

Dissolve 100 mg of Coomasie Brilliant Blue G-250 in 50 ml of 95% ethanol. Add 100 ml of 85% phosphoric acid, and bring to a final volume of 1 litre with distilled water.

Phosphoric acid is extremely corrosive. Handle with care.

Bromophenol blue (Sodium salt, MW 692.0)

0.001 (w/v)

Dissolve 1 mg of Bromophenol blue, sodium salt to a final volume of 100 ml with either water or buffer.

n-Butanol (C_4H_9OH MW 74.12)

Density = 0.8098 grams/ml 1.1 M

The butanol can be weighed (81.5 grams) or measured volumetrically by using the density. That is, 81.5 grams 0.8098 grams/ml or 100.7 ml of n-butanol. Weigh or measure the appropriate amount and dilute to a final volume of 1 litre with water.

Calcium chloride (MW 110.99)

0.001 M

Dissolve 0.111 grams of anhydrous calcium chloride to a final volume of 1 litre with water or buffer.

0.08 M

Dissolve 8.879 grams of anhydrous calcium chloride to a final volume of 1 litre with water or buffer.

2% (w/v)

Dissolve 2 grams of anhydrous calcium chloride to a final volume of 100 ml with water or buffer.

cAMP (Adenosine monophosphate, cyclic MW 329.22)

0.001 M (1mM)

Dissolve 33 mg of cAMP to a final volume of 100 ml with water, buffer or media.

Carnoy fixative

Combine 10.0 ml of glacial acetic acid with 60.0 ml of absolute ethyl alcohol and 30.0 ml of chloroform.

Chloroplast homogenization buffer

To 400 ml of distilled water, add 30.058 grams of sorbitol, 2.23 grams of sodium pyrophosphate, 0.407 grams of magnesium chloride, and 0.176 grams of ascorbic acid. Adjust the pH to 6.5 with HCl and dilute to a final volume of 500 ml.

Chloroplast suspension buffer

To 400 ml of distilled water, add 30.058 grams of sorbitol, 0.372 grams of EDTA, 0.102 grams of magnesium chloride and 5.958 grams of HEPES buffer. Adjust the pH to 7.6 with NaOH and dilute to a final volume of 500 ml.

Citric acid ($H_3C_6H_5O_7 \cdot H_2O$ MW 210.14)

0.1 M

Dissolve 21.01 grams of citric acid to a final volume of 1 litre.

Citrate buffer (Sodium phosphate-Citrate buffer)

0.001 M

PH 4.8

Add 493 ml of 0.2 M Na_2HPO_4 to 507 ml of 0.1 M citric acid.

PH 3.6

Add 322 ml of 0.2 M Na_2HPO_4 to 678 ml of 0.1 M citric acid.

PH 4.2

Add 414 ml of 0.2 M Na_2HPO_4 to 586 ml of 0.1 M citric acid.

PH 5.4

Add 557.5 ml of 0.2 M Na_2HPO_4 to 442.6 ml of 0.1 M citric acid.

PH 6.0

Add 631.5 ml of 0.2 M Na_2HPO_4 to 368.5 ml of 0.1 M citric acid.

PH 6.6

Add 727.5 ml of 0.2 M Na_2HPO_4 to 272.5 ml of 0.1 M citric acid.

PH 7.2

Add 869.5 ml of 0.2 M Na_2HPO_4 to 130.5 ml of 0.1 M citric acid.

PH 7.8

Add 957.5 ml of 0.2 M Na_2HPO_4 to 42.5 ml of 0.1 M citric acid.

Cobaltous Nitrate (MW 182.96)

2% (w/v)

Dissolve 2.0 grams of cobaltous nitrate (hexahydrate is very soluble) to a final volume of 100 ml with water. Keep well closed in a cool place.

Coomasie blue (Coomasie Brilliant Blue R250)

0.25% (w/v)

Dissolve 2.50 grams of Coomasie Brilliant Blue R250 to a final volume of 1 litre with 20% (w/v) trichloroacetic acid (TCA).

Copper tartrate/carbonate (CTC)

Dissolve 0.5 grams of copper sulfate and 1.0 gram of potassium sodium tartrate to a final volume of 100 ml with water. Combine 1.0 ml of this solution with 50 ml of 2% Na_2CO_3 in 0.1 N NaOH. Must be made fresh, prior to use.

Crystal Violet

Dissolve 0.1 grams of crystal violet and 0.25 ml of glacial acetic acid to a final volume of 100 ml with water.

Dische diphenylamine reagent

Dissolve 500 mg of diphenylamine in 49 ml of glacial acetic acid. Add 1.0 ml of concentrated HCl.

Dithiothreitol (Cleland's Reagent MW 154.3)

0.01 M

Dissolve 154 mg of dithiothreitol to a final volume of 100 ml with water or buffer.

DOPA (3-(3,4-Dihydroxyphenyl)-L-alanine MW 197.19)

8 mM

Dissolve 158 mg of L-DOPA to a final volume of 100 ml with water or buffer. Note that the maximum solubility of DOPA in water is 165 mg/100 ml (8.3 mM).

EDTA (Ethylenediaminetetraacetic acid MW 292.24)

1 M

Dissolve 292.24 grams of EDTA, free acid to a final volume of 1 litre. If the more soluble disodium salt of EDTA is used, adjust the weight accordingly. The pH can be adjusted with acetic acid or NaOH. For corresponding concentration dilutions, multiply the weight in grams by the desired molarity. For example, for 10 mM EDTA, multiply 292.24 X 0.010 to obtain 2.92 grams of EDTA per litre.

Ethanol-acetic acid fixative for histology (3:1)

To 75 ml of absolute alcohol, add 25 ml of glacial acetic acid. Must be made fresh, just prior to use.

Fetal Calf Serum (FCS)

While it is possible to prepare your own serum from whole blood, it is easier (and safer) to purchase FCS from a reputable supplier. Commercial sources are free of mycoplasma, pre-sterilized and controlled for the presence of antibodies. There are a number of serum substitutes available on the market and these may be less expensive when storage is considered.

Folin-Ciocalteu Reagent

This is usually purchased premixed, since it is difficult to make. Also known as 2N Folin-phenol reagent.

Giemsa stain

Prepare a stock solution by dissolving 3.8 grams of giemsa powder in 25 ml of glycerin. Heat gently with stirring for about 2 hours at 60 ° C. Cool and add 75 ml of methanol (neutral, acetone free).

For a working solution, dilute the stock solution 1/10 with water before use.

For chromosome banding, combine 5.0 ml of stock Giemsa, 3.0 ml of absolute methanol, 3.0 ml of 0.1 M citric acid and 89 ml of distilled water. Adjust the pH of the solution to 6.6 with Na_2HPO_4.

Glutaraldehyde (GTA)

5 %

GTA is usually supplied as a 25% or 50% (w/v) solution. It is used for electron microscope fixation as a 5% solution in a buffer. For routine use, add 20 ml of 25% GTA to 80 ml of 0.2 M sodium cacodylate buffer, pH 7.4.

Gram's iodine

Dissolve 0.33 grams of iodine and 0.67 grams of potassium iodide to a final volume of 100 ml with water.

HEPES (N-[2-Hydroxyethyl]piperazine-N'-[2-ethanesulfonic acid] MW 238.3)

50 mM

Dissolve 11.92 grams of HEPES, free acid to a final volume of 1 litre. If hemisodium salt is used, adjust weight accordingly. Do not use sodium salts unless specified. Hemisodium salt contains 0.5 moles of sodium for each mole of HEPES.

10 mM
pH7.6

Dissolve 2.38 grams of HEPES, free acid in 900 ml of water. Adjust the pH with NaOH or HCl to 7.6. Adjust the final volume to 1 litre with distilled water.

Hydrochloric Acid (HCl MW 36.46)

Concentrated HCl has a molarity of approximately 11.6.

HCl is a gas, which is soluble in water and which comes in the form of concentrated reagent grade HCl. This solution is approximately 36-38% (w/v) HCl. To make a 1 N solution, add 86 ml of concentrated HCl to 800 ml of water and dilute to a final volume of 1 litre. For 0.1 N, dilute the 1 N by a factor of 10.

For % solutions, note that liquid HCl is only 38% HCl, thus a 1% solution would require 2.6 ml of concentrated HCl (1/.38) per final volume of 100 ml.

Krebs Phosphate Ringers (KPR)

Prepare each of the following separately:

0.90% (w/v) NaCl

1.15% (w/v) KCl

1.22% (w/v) $CaCl_2$

3.82% (w/v) $MgSO_4 \cdot H_2O$

0.1 M phosphate buffer, pH 7.4 (17.8 grams of $Na_2HPO_4 \cdot H_2O$
+ 20 ml of 1 N HCl, diluted to 1 litre)

To mix, combine 200 ml of NaCl, 8 ml of KCl, 6 ml of $CaCl_2$ and 2 ml of $MgSO_4$. Carefully, and with constant stirring, add 40 ml of phosphate buffer.

LPS buffer (Lower Pad Solution buffer)

Dissolve 1.5 grams of KCl, 0.5 grams of $MgCl_2$, and 0.5 grams of streptomycin sulphate in 500 ml of water. Add 40 ml of 1 M phosphate buffer, pH 6.5 and dilute to 1 litre with water.

Mayer's hematoxylin

A. Dissolve 1.0 gram of hematoxylin in 10 ml of absolute ethanol. Dissolve 20 grams of potassium alum (KAl(SO_4)_2 · 12H_2O) in 200 ml of water. In a chemical hood, with protection against explosion, bring the potassium alum solution to a boil and add hematoxylin/ethanol mixture. Continue to boil for approximately 1 minute. Add 0. 5 grams of mercuric oxide and cool rapidly. Add 0.5 ml of glacial acetic acid. Filter before use. This mixture is stable for about two months.

B. Alternatively: Dissolve 5.0 grams of hematoxylin in 50 ml of absolute ethanol and add to 650 ml of warm water. Heat gently until the hematoxylin dissolves and then add 300 ml of glycerin, 0.3 grams of sodium iodate and 20 ml. of glacial acetic acid. Co ol and make volume up to 1 litre with distilled water. Filter before use.

MES (2-(N-Morphilino)ethanesulfonic acid MW 195.2)

0.1 M

Dissolve 1.952 grams of MES to a final volume of 100 ml with water or buffer.

Methanol/Acetic Acid (for fixing proteins in acrylamide gels)

45%:12%

Add 120 ml of glacial acetic acid to 450 ml of methanol and dilute to a final volume of 1 litre with water.

Methanol/Acetic Acid (for destaining or fixing proteins in acrylamide gels)

5%:7%

Add 70 ml of glacial acetic acid to 50 ml of methanol and dilute to a final volume of 1 litre with water.

Methyl green

0.2% (w/v)

Dissolve 0.2 grams of methyl green to a final volume of 100 ml with 0.1 M acetate buffer, pH 4.2.

Acetate buffer (0.1 M pH 4.2) is prepared by dissolving .361 grams of sodium acetate (trihydrate) in approximately 80 ml of water. Add .42 ml of glacial acetic acid and adjust the volume to 100 ml with water.

Microtubule buffer (MT buffer)

Dissolve 19.52 grams of MES in 800 ml of distilled water. Add 0.380 grams of EGTA and 47.62 grams of $MgCl_2$. Adjust the pH to 6.4 with HCl or NaOH and dilute to a final volume of 1 litre with disti lled water.

Periodic Acid (Periodate Used for PAS reaction)

Dissolve 0.6 grams of periodic acid in 100 ml of water and add 0.3 ml of concentrated nitric acid.

Phenol mixture

Combine 555 ml of aqueous phenol (or 500 grams of phenol crystals plus 55 ml of water) with 70 ml of mcresol. Add 0.5 grams of 8-hydroxyquinoline.

Phenol will cause severe burns and readily dissolves all plastic and rubber compounds. Use extreme caution when handling this compound.

Phosphate buffered saline (PBS)

Mix 100 ml Ca^{++}/Mg^{++} free 10X PBSA with 800 ml of distilled water. Separately, dissolve 0.1 gram of magnesium chloride and 0.1 gram of anhydrous calcium chloride to a final volume of 100 ml with water. With constant stirring, slowly add the magnesium/calcium chloride solution to the diluted PBSA. If a precipitate forms, start over, and add slower with continuous stirring.

Ca^{++}/Mg^{++} free Phosphate buffered saline - 10X (10X PBSA)

Dissolve 80 grams of NaCl, 2.0 grams of KCl, 15.0 grams of Dibasic sodium phosphate and 2.0 grams of Monobasic potassium phosphate in 1 litre of distilled water. This makes a 10X solution of Ca^{++}/Mg^{++} free phosphate buffered saline. Dilute 1:10 prior to use. Store in a refrigerator.

Phosphate buffered saline-Tween 20 (4.8)

Mix PBS and add 0.1% (v/v) Tween 20.

Phytohemaglutinin (PHA)

Available as kidney bean lectin. It is typically used as a stock solution of 10-20 g/ml in balanced salt solution. For tissue culture it must be cold sterilized prior to use.

Protein Buffer

Dissolve 1.46 grams of KH_2PO_4 and 0.92 grams of K_2HPO_4 in 80 ml of distilled water. Add 2.5 grams of crystalline serum albumin and adjust the volume to a final 100 ml with water.

Safranin

Dissolve 2.5 grams of Safranin O in 10 ml of 95% ethanol and dilute to 100 ml with water.

Saline (NaCl)

0.85% (w/v)

Saline refers to a solution of NaCl, with the most common usage for that which is isotonic to mammalian blood cells, notable a 0.85% or 0.9% solution. To mix, dissolve 8.5 grams of NaCl to a final volume of 1 litre with water.

Saline citrate (1/10 dilution of SSC)

Dissolve 0.878 grams of NaCl and 0.294 grams of sodium citrate to a final volume of 1 litre with water.

Saline citrate buffer (SSC)

20X

It is common to prepare this buffer as a stock 20X solution, to be diluted to 2X, 1X or 0.1X prior to use. To prepare a 20X stock solution, dissolve 175 grams of NaCl and 88 grams of sodium citrate in 900 ml of water. Adjust the pH to 7.0 with 1 N HCl and bring to a final volume of 1 litre.

For use, as a 1X SSC, dilute 1 part 20X stock with 19 parts distilled water. For a 2X SSC, dilute 1 part 20X stock with 9 parts water.

Schiff's Reagent

Dissolve 0.8 grams of basic fuchsin in 85 ml of distilled water. Add 1.9 grams of sodium metabisulfite and 15.0 ml of 1 N HCl. Place the solution in separatory funnel and shake at 2 hour intervals for a period of approximately 12 hours. Add 200 grams of activated charcoal, shake for 1 minute and filter the clear solution. If the solution is still pink, add another 100 grams of charcoal and shake for an additional minute. Filter and store in a dark bottle. Solution should be clear (no pink coloration) for use.

Scott solution

Dissolve 2.0 grams of sodium bicarbonate and 20.0 grams of magnesium sulfate in water to a final volume of 1 litre. Add a pinch of thymol to retard the growth of molds.

1X SDS-Electrophoresis Running Buffer

Dilute 5X Tris-Glycine buffer to 1X and add 1.0 gram of SDS per litre of 1X Tris-Glycine. The pH should be 8.3 after dilution.

2X SDS Sample Buffer

Dissolve 1.52 grams of Tris base, 2.0 grams of SDS, 20 ml of glycerin, 2.0 ml of -mercaptoethanol and 1 mg of bromophenol blue to a final volume of 100 ml with water.

Siliconized pipettes

Pasteur pipettes can be siliconized by soaking them in a beaker containing 5% (v/v) dichlorodimethylsilane in chloroform for about 1 minute. Remove, drain and rinse several times with distilled water. Bake the pipettes at 180° C for 2 hours and cool before use.

Dichlorodimethylsilane and chloroform are both toxic and volatile. Use only in proper fume hood and keep all flames away from work area. Insure that all silicone and chloroform are removed from glassware before placing in an oven.

Sodium acetate buffer

1 M pH 5.7

To 925 ml of 1 M sodium acetate, add 75 ml of 1 M acetic acid.

Sodium cacodylate buffer

0.2 M pH 7.4

Prepare a 0.2 M solution of cacodylic acid, sodium salt (MW 159.91). Dissolve 3.20 grams of cacodylic acid, sodium salt to a final volume of 100 ml with water. Adjust the pH to 7.4 with HCl.

Cacodylic acid contains arsenic. Handle properly.

Sodium citrate/formaldehyde (for silver stained proteins)

Dissolve 5.0 grams of sodium citrate in 800 ml of water, add 5.0 ml of concentrated formalin (37% formaldehyde solution) and dilute to 1 litre with water.

Sodium phosphate buffer

These are the most common buffers used in biology. They are produced by adding equimolar solutions of KH_2PO_4 and Na_2HPO_4. Equal volumes of the two will yield a pH of 7.0, while sodium phosphate will increase the pH. Increased volumes of potassium phosphate will decrease the pH. The pH can be adjusted from 5.4 to 8.2.

If a pH of 7.0-8.2 is desired, start with 500 ml of sodium phosphate and add potassium phosphate while stirring and monitoring the pH with a pH meter until the desired pH is reached.

If a pH of 5.4-7.0 is desired, start with 500 ml of potassium phosphate and add sodium phosphate until the desired pH is reached.

Typically, the molarity of the buffer will range from 0.01 to 0.1 M. Use the appropriate molarity of KH_2PO_4

and Na_2HPO_4. That is, if 0.05 M buffer is desired, use 0.5 M KH_2PO_4 and 0.5 M Na_2HPO_4 as directed above.

Sorenson phosphate buffer
sodium phosphate buffer. 0.2 M pH 7.5

Dissolve 24.14 grams of Na_2HPO_4 and 4.08 grams of KH_2PO_4 in 800 ml of water. Dilute to a final volume of 1 litre.

Sulfurous acid (for Feulgen Reaction)

Add 1.0 ml of concentrated HCl and 0.4 grams of sodium bisulfite to 100 ml of distilled water. This solution should be made fresh prior to use. It does not store well.

Swabbing detergent

For tissue culture purposes, use a non-toxic detergent designed for surgical scrubbing. e.g. Phisohex, Betadine or equivalent. For most routine swabbing, 70% (v/v) ethanol is sufficient and has the advantage that it will leave no residue.

Toluidine blue

0.1% (w/v)

Dissolve 0.1 grams of toluidine blue in 10 ml of ethanol and add water or citrate buffer (pH 6.8-7.2) to a final volume of 100 ml.

Tris buffer

There are many variations on the basic Tris-HCl buffer combination, most of which are commercially available. Solutions with EDTA are known as TE buffers, while solutions with EDTA and acetic acid are known as TAE buffers. The terminology varies with the author, with Tris buffer being used to mean Tris-HCl solutions.

The basic buffer is a combination of Tris (tris(h ydroxymethyl)aminomethane) and HCl acid. These are sometimes referred to as Tris-base and Tris-HCl solutions. Tris buffers should not be used below a pH of 7.2 or above a pH of 9.0. Tris buffers are also extremely temperature senstive. Directions are given for room temperature (25 ° C). The pH will decrease approximately 0.028 units for each degree decrease in temperature.

1 M

Dissolve 121 grams of Tris in 800 ml of distilled water. Adjust the pH with concentrated HCl. Dilute to a final volume of 1 litre. Lower required molarities can be diluted from this stock or mixed as combinations of lower molarities of Tris and HCl. It is important to measure the pH at the temperature an d molarity that will be used in the final analysis.

Tri-Glycine buffer

5X

Dissolve 15.1 grams of tris base and 72.0 grams of glycine to a final volume of 1 litre. For use, dilute 1 part 5X buffer with 4 parts water.

Trypsin

0.25%

Dissolve 0.25 grams of crude trypsin in PBSA to a final volume of 100 ml. Cold sterilize by filtration.

Trypticase soy broth

Add 17.0 grams of trypticase peptone, 3.0 grams of phytone peptone, 5.0 grams of sodium chloride, 2.5 grams of dipotassium phosphate and 2.5 grams of glucose to 1 litre of water. Adjust the pH to 7.3, and autoclave.

Tween 20 or 80 (Polyoxyethylene sorbitan mono-oleate)

1% (v/v)

Add 1.0 ml of Tween to 90 ml of water. Mix and dilute to a final volume of 100 ml with water. Note that Tween is extremely viscous and care must be taken to accurately pipette 1.0 ml. Wipe the outside of the pipette before dispensing.

Uranyl acetate (MW 424.19)

5% (w/v)

Dissolve 5.0 grams of uranyl acetate to a final volume of 100 ml in 50% (v/v) ethanol. Store in the dark at room temperature. Allow at least 24 hours for the uranyl acetate to completely dissolve. This solution will keep for about 3 months.

17

Waste Disposal

It is always important to dispose off chemicals after the experiments are over as it is dangerous to keep them exposed for long time.

18.1 Acids

All waste acidic chemicals should be neutralized with soda ash or lime or caustic soda solution before throwing them into drain.

18.2 Alkalies

All alkaline chemicals should be neutralized using acids before disposing them off into drain.

18.3 Toxic Chemicals

No toxic chemical should be drained before detoxifying its effect. They should be disposed off using specified procedure given in the laboratory operating manuals.

Appendix-I

Approximate Minimum Water Activity (a^W)

Organisms	water activity a^w
Groups	
Major spoilage bacteria	*0.90*
Major spoilage yeasts	*0.88*
Major spoilage molds	*0.80*
Major Halophilic bacteria	*0.75*
Major Xerophilic molds	*0.61*
Major Oxmophilic yeasts	*0.61*
Microorganisms (Specific)	
Clostridium botulinum, type E	*0.97*
Pseudomonasspp.	*0.97*
Acinetobacterspp.	*0.96*

Organisms	**water activity a^w**
Escherichia coli	*0.96*
Enterobacter aerogenes	*0.95*
Bacillus subtilis	*0.95*
Clostridium bot.,types A and B	*0.94*
Candida utilis	*0.94*
Vibrio parahaemolyticus	*0.94*
Eurotium repens	*0.72*
Rhizopus stolonifer	*0.93*
Mucor spinosus	*0.93*
Zygosaccharomyces rouxii	*0.62*
Candida scottii	*0.92*
Trichosporon pullulans	*0.91*
Candida zeylanoides	*0.90*
Geotrichum candidum	*0.90*
Trichotheciumspp.	*0.90*
Staphylococcus aureus	*0.86*
Alternaria citri	*0.84*
Pencillium patulum	*0.81*
Aspergillus conicus	*0.70*
Aspergillus echinulatus	*0.64*
Xeromyces bisporus	*0.51*

Appendix-II

Cluster of Differentiation Molecules (CDs)

CD Molecule	Function
CD 2	The CD2 molecule is a glycoprotein present on T cells, thymocytes and NK cells. It functions as an intercellular adhesion molecule and binds to the human leukocyte function-associated antigen-3 or LFA-3 (CD58). CD2 may be involved in the attachment of THcells to antigen- presenting cells and cytotoxic T cells to target cells, and in other cellular interactions. It also is a signal-transducing molecule and may help stimulate T cells to proliferate and secrete cytokines.
CD 4	The CD4 molecule is a glycoprotein found primarily on T cells associated with helper/inducer regulatory functions and to a lesser degree on monocytes and macrophages. It is a cell adhesion molecule with great affinity for class II MHC.

CD Molecule	Function
CD 8	The CD8 molecule is a polypeptide found on the surface of cytotoxic T cells. It is a molecule that participates in the interaction of class I MHC found on target cells.
CD 23	CD23 is a protein found on the surface of IgM-bearing B cells, eosinophils, macrophages, and platelets. Soluble CD23 is a B-cell growth factor and upregulates IgE synthesis in conjunction with interleukin-4 released from T cells.
CD 35	CD35 is a glycoprotein present on erythrocytes, PMNs, monocytes, all B and some T cells, mast cells, and in soluble form in the plasma. It is the receptor for complement C3b/C4b. The functions of CD35 vary according to the cell type possessing the receptor. For example, CD35 on erythrocytes binds immune complexes bearing C3b or C4b and promotes their clearance from the circulation. On B cells, CD35 augments the maturation of these cells into antibody-secreting cells. It also aids leukocytes in binding and phagocytosing C3b- or C4b-coated objects.
CD 54	CD54, also known as intercellular adhesion molecule-1 (ICAM-1), is a protein that is an important early marker of immune activation and response. It is a ligand for the lymphocyte function-associated antigen-1 (LFA-1) found on lymphocytes, monocytes, and neutrophils. The CD54/LFA-1 interaction is important in T-cell-antigen-specific responses and lymphocyte emigration into inflammatory sites. CD54 also is found on mucosal epithelial cells and erythrocytes. Recent evidence suggests CD54 is the cellular receptor for the human rhinovirus and Plasmodium falciparum-infected erythrocytes.

Appendix-III

Common Buffer Compounds

Name	pKa (25 C)	Buffer Range	Mol.Wt	Full Compound Name
TAPS	8.43	7.7-9.1	243.3	3-[tris (hydroxym ethyl) methyl] amino} propanesulfonic acid
Bicine	8.35	7.6-9.0	163.2	N, N-bis (2-hydroxyethyl) glycine
Tris	8.06	7.5-9.0	121.14	tris (hydroxymethyl) methylamine
Tricine	8.05	7.4-8.8	179.2	N-tris (hydroxymethyl) methylglycine
HEPES	7.55	6.8-8.2	238.3	4-2-hydroxyethyl-1-piperazineethane sulfonic acid

Name	pKa (25 C)	Buffer Range	Mol.Wt	Full Compound Name
TES	7.40	6.8-8.2	229.20	2-{[tris (hydroxymethyl) methyl]amino} ethanesulfonic acid
MOPS	7.20	6.5-7.9	209.3	3-(N-morpholino) propanesulfonic acid
PIPES	6.76	6.1-7.5	302.4	piperazine-N, N'-bis (2-ethanesulfonic acid)
Cacodylate	6.27	5.0-7.4	138.0	dimethylarsinic acid
SSC	7.0	6.5-7.5	189.1	saline sodium citrate
MES	6.15	5.5-6.7	195.2	2-(N-morpholino) ethanesulfonic acid

Appendix-IV

Comparison of Procaryotic and Eucaryotic Cells

Property	Procaryotes	Eucaryotes
Organization of Genetic Material		
True membrane-bound nucleus	Absent	Present
DNA complexed with histones	No	Yes
Number of chromosomes	One[a]	More than one
Introns in genes	Rare	Common
Nucleolus	Absent	Present
Mitosis occurs	No	Yes
Genetic Recombination	Partial, unidirectional transfer of DNA gametes	Meiosis and fusion of

Property	Procaryotes	Eucaryotes
Mitochondria	Absent	Present
Chloroplasts	Absent	Present
Plasma Membrane with Sterols	Usually no[b]	Yes
Flagella	Submicrosopic in size; composed of one fiber	Microscopic in size; membrane bound; usually 20 microtubules in 9 + 2 pattern
Endoplasmic Reticulum	Absent	Present
Golgi Apparatus	Absent	Present
Cell Walls	Usually chemically complex with peptidoglycan	Chemically simpler and lacking peptidoglycan[c]
Differences in Simpler Organelles		
Ribosomes	70S	80S (except in mitochondria and chloroplasts)
Lysosomes and peroxisomes	Absent	Present
Microtubules	Absent or rare	Present
Cytoskeleton	May be absent	Present
Differentiation	Rudimentary	Tissues and organs

[a]Plasmids may provide additional genetic information; [b]Only the mycoplasmas and methanotrophs (methane utilizers) contain sterols. The mycoplasmas cannot synthesize sterols and require them preformed. Many procaryotes contain hopanoids; [c]The mycoplasmas and Archaea do not have peptidoglycan cell walls.

Appendix-V

Conversion Factors

From	To	Multiply by
Acre (Ac)	hectare (ha)	0.4047
British thermal unit(btu)	Joule (J)	1055.056
Btu per hour (btu/hr)	watt (w)	0.2930711
Btu per lb (btu/lb)	calories per gram (cal/g)	0.55556
Btu	Joule	1055.1
Btu/degree	F×hr×ft2 w/m2 - degree k	5.678263
Calorie (cal)	Joule (J)	4.1868e
Cubic foot (cu ft)	cubic meter (cu m)	0.02831685
Cubic foot	litre (L)	28.317
Cubic inch (cu in)	cubic meter (cu m)	0.00001639
Cubic inch	cubic centimetre (cu cm)	16.387
Cubic yard (cu yd)	cubic meter (cu m)	0.7645549
Darcy	centimeter per second (cm/sec)	0.000968

From	To	Multiply by
Degree Fahrenheit (f)	degree celsius (°C)	tc = (tf-32)/ 1.8
Degree Fahrenheit (f)	kelvin (K)	tk = (tf+459.7)/ 1.8
Dram (avoirdupois)	gram	1.7718
Feet per day (ft/day)	centimeter per second (cm/sec)	0.000352
Fluid ounce (fl oz)	cubic meter (cu m)	0.00002957
Fluid ounce (fl oz)	milliliters (ml)	29.57353
Foot (ft)	meter (m)	0.3048
Foot	centimetre (cm)	30.48
Gallon (gal)	cubic meter (cu m)	0.00378541
Gallon (U.S., liquid)	liter (L)	3.7854
Grain	kilogram (Kg)	0.0000648
Grain	milligram (mg)	64.799
Horsepower (hp) (550 ft-lb/sec)	watt (W)	745.70
Horsepower	Joule/second	745.70
Inch	meter (m)	0.0254
Inch	millimeter (mm)	25.4
Inch of h2o	pascal	249.09
Inch of hg	pascal	3386.4
Inch	centimetre (cm)	2.54
Kelvin (K)	degree celsius (°C)	tc=tk-273.15
Kilowatt-hour (KWh)	Joule (J)	3,600,000

From	To	Multiply by
Kip (1000 lb)	kilogram (Kg)	453.6
Kip (1000 lb)	Newton (N)	4,448.222
Kip per linear foot (klf)	kilogram per meter (Kg/m)	0.001488
Kip per square inch (ksi)	megapascal (MPa)	6.894757
Mile (U.S. statute)	kilometer (Km)	1.609347
Mile per hour (mph)	kilometer per hour(Km/hr)	1.609347
Mile per hour (mph)	meter per second (m/s)	0.44704
Ounce (avoirdupois)	gram (g)	28.350
Ounce (U.S., liquid)	millilitre (mm)	29.574
Pound (avoirdupois)	kilogram (Kg)	0.45359
Pound (lb)	Newton (N)	4.448222
Pound per cub yd (lb/cu yd)	kilogram per cub m (Kg/cu m)	0.5933
Pound per cubic foot (pcf)	kilogram per cub m (Kg/cu m)	16.01846
Pound per linear foot (plf)	kilogram per meter (Kg/m)	1.488
Pound per square foot (psf)	Kg per square meter (Kg/sq m)	4.8824
Pound per square foot (psf)	pascal (Pa)	47.88
Pound per square inch (psi)	megapascal (MPa)	0.00689476
Pound per square inch (psi)	pascal (Pa)	6,894.757

From	**To**	**Multiply by**
Quart (U.S., liquid)	liter (L)	0.94635
Square foot	square meter (sq m)	0.092903
Square inch (sq in)	square meter (sq m)	0.00064516
Square inch	square centimetre (sq cm)	6.4516
Square yard	square meter (sq m)	0.83613
Ton, 2000 lb	kilogram (Kg)	907.1848
Yard	meter (m)	0.9144

Appendix-VI

Cytokines: Immune Responses

Cytokine	Cell Source	Functions
IL-1 (interleukin-1)	Monocytes/macrophages, endothelial cells, fibroblasts, neuronal cells, glial cells, keratinocytes, epithelial cells	Produces a wide variety of effects on the differentiation and function of cells involved in inflammatory and immune responses; also affects central nervous and endocrine systems; it is an endogenous pyrogen
IL-2 (interleukin-2, T-cell growth factor)	T cells (TH1)	Stimulates T-cell proliferation and differentiation, enhances cytolytic activity of NK cells, promotes proliferation and immunoglobin secretion of activated B cells

Cytokine	Cell Source	Functions
IL-3 (interleukin-3)	T cells, keratinocytes, neuronal cells, mast cells	Stimulates the production and differentiation of macrophages, neutrophils, eosinophils, basophils, mast cells
IL-4 (interleukin-4, B-cell growth factor-1 [BCGF-1], B-cell stimulatory factor-1 [BCSF-1])	T cells (TH2), macrophages, mast cells, basophils, B cells	Induces the differentiation of naive $CD4^+$ T cells into T-helper-like cells; induces the proliferation and differentiation of B cells; exhibits diverse effects on T cells, monocytes, granulocytes, fibroblasts, endothelial cells
IL-5 (interleukin-5)	T cells (TH2)	Growth and activation of B cells and eosinophils; activation of eosinophil function; chemotactic for eosinophils
IL-6 (interleukin-6, cytotoxic T-cell differentiation factor, B-cell differentiation factor)	TH2 cells, monocytes/ macrophages, fibroblasts, hepatocytes, endothelial cells, neuronal cells	Activates hematopoietic cells; induces growth of T cells, B cells hepatocytes, keratinocytes, and nerve cells; stimulates the production of acute-phase proteins
IL-8 (interleukin-8)	Monocytes, endothelial cells, fibroblasts, alveolar	Chemoattractant for PMNs; causes degranulation and expression of receptors; inhibits adhesion of

Cytokine	Cell Source	Functions
	epithelium, T cells, keratinocytes, neutrophils,	PMNs to endothelium; hepatocytes promotes migration of PMNs through endothelium; induces adhesion of neutrophils to endothelial cells
IL-10 (interleukin-10)	T cells (TH2), B cells, macrophages, keratinocytes	Reduces the production of IFN-γ, IL-1, TNF-α, and IL-6 by macrophages; in combination with IL-3 and IL-4, causes mast cell growth; in combination with IL-2, causes growth of cytotoxic T cells and differentiation of $CD8^+$ cells
IFNs α/β **(interferons**α/β**)**	T cells, B cells, monocytes/ macrophages, fibroblasts	Antiviral activity, stimulates macrophage activity, modulates MHC class I and II protein expression on various cells, regulates the development of the specific immune response
IFN-γ (interferon-γ)	T cells (TH1, TC), NK cells	Activation of T cells, macrophages, neutrophils, and NK cells; increases class I and II MHC molecules

Cytokine	Cell Source	Functions
TNF-α (tumor necrosis factor-α[cachectin])	T cells, macrophages and NK cells	A wide variety of effects due to its ability to mediate expression of genes for growth factors and cytokines, transcription factors, receptors, inflammatory mediators, and acute-phase proteins; plays a role in host resistance to infection byserving as an immunostimulant and mediator of the inflammatory response; cytotoxic for tumor cells
TNF-β (tumor necrosis factor-β [lymphotoxin])	T cells, B cells	Same as TNF-α
G-CSF (granulocyte colony-stimulating factor)	T cells, macrophages, neutrophils	Enhances the differentiation and activation of neutrophils
M-CSF (macrophage colony- stimulating factor)	T cells, neutrophils, macrophages, fibroblasts, endothelial cells	Stimulates various functions of monocytes and macrophages, promotes the growth and development of macrophage colonies from undifferentiated precursors

Appendix-VII

Differences in Spoilage Processes in Relation to Food Characteristics

Food	Substrate	Chemical Reactions or Processesa	Typical Products and Effects
Fruits	Pectin	Pectinolysis	Methanol, uronic acids (loss of fruit structure, soft rots)
Meat	Proteins	Proteolysis, deamination	Amino acids, peptides, amines, H_2S, ammonia, Sindole (bitterness, souring, bad odor, sliminess)
Starchy foods	Carbohydrates	Hydrolysis, fermentations	Organic acids, CO_2, mixed alcohols (souring, acidification)
Butter/ Ghee	Lipids	Hydrolysis, fatty acid degradation	Glycerol and mixed fatty acids (rancidity, bitterness)

[a]Other reactions also occur during the spoilage of these substrates.

Appendix-VIII

Enterobacteriaceae: Some Characteristics of Selected Genera

Characteristics	Escherichia	Shigella	Salmonella	Citrobacter	Proteus	Yersinia	Klebsiella	Enterobacter	Erwinia	Serratia
Methyl red	+	+	+	+	+	+	(+)[a]	(-)[b]	+	d[c]
Voges-Proskauer	-	-	-	-	d	- (37°C)	(+)	+	(+)	+
Indole production	(+)	d	-	d	d	d	d	-	(-)	(-)
Citrate use	-	-	(+)	+	d	(-)	(+)	+	(+)	+
H2S production	-	-	(+)	d	(+)	-	-	- (	+)	-
Urease	-	-	-	(+)	+	d	(+)	(-)	-	-
?-galactosidase	(+)	d	d	+	-	+	(+)	+	+	+

Characteristics	Escherichia	Shigella	Salmonella	Citrobacter	Proteus	Yersinia	Klebsiella	Enterobacter	Erwinia	Serratia
Gas from glucose	+	-	(+)	+	+	(-)	(+)	(+)	(-)	d
Acid from lactose	+	-	(-)	d	-	(-)	(+)	(+)	d	d
Phenylalanine deaminase	-	-	-	-	+	-	-	(-)	(-)	-
Lysine decarboxylase	(+)	-	(+)	-	-	(-)	(+)	d	-	d
Ornithine decarboxylase	(+)	d	(+)	(+)	d	d	-	(+)	-	d
Motility	d	-	(+)	+	+	- (37°C)	-	+	+	+
Gelatin liquifaction (22°C)	-	-	-	-	+	(-)	-	d	d	(+)
% G + C	48-52	49-53	50-53	50-52	38-41	46-50	53-58	52-60	50-58	53-59

[a](+) usually present; [b](-) usually absent; [c]d, strains or species vary in possession of characteristic

Appendix-IX

Food Additives: E-Codes

Code	Chemical Name	Description	Remarks
100	Curcumin (C.I. 75300)	Colouring	Orange-yellow colouring extracted from the rhizome of a plant of the ginger family called Curcuma longa (turmeric)
101	Riboflavin/Lactofavin/ Vitamin B2*	Colouring/ Vitamin	Yellow/orange-yellow manufactured from yeast/other fermenting organisms. May also be synthesised from xylene, ribose or alloxan
102	Tartrazine/FD&C Yellow 5 (C.I. 19140)	Colouring	Synthetic yellow colouring
104	Quinoline Yellow (C.I. 47005)	Colouring	Synthetic dull yellow colouring
107	Yellow 2G (C.I. 18965)	Colouring	Synthetic yellow colouring
110	Sunset Yellow FCF/FD&C Yellow 6(C.I. 15985)	Colouring	Synthetic yellow colouring

Code	Chemical Name	Description	Remarks
120	Cochineal/Carminic acid/ Carmines (C.I. 75470) *	Colouring	Natural red colouring which may be extracted from the bodies of female insects called Dactilopius coccus
122	Carmoisine/Azorubine (C.I. 14720)	Colouring	Synthetic red colouring
123	Amaranth/FD&C Red 2 (C.I. 16185)	Colouring	Synthetic purplish red colouring
124	Ponceau 4R/Cochineal Red A (C.I. 16255)	Colouring	Synthetic red colouring
127	Erythrosine/FD&C Red 3 (C.I. 45430)	Colouring	Synthetic cherry pink to red colouring
128	Red 2G (C.I. 18050)	Colouring	Synthetic red colouring
129	Allura Red AC/Food Red 17/FD&C Red 40 (C.I. 16035)	Colouring	Synthetic red colouring
131	Patent Blue V (C.I. 42051)	Colouring	Synthetic dark bluish-violet colour
132	Indigo Carmine/Indigotine/ FD&C Blue 2 (C.I. 73015)	Colouring	Synthetic blue colouring

Code	Chemical Name	Description	Remarks
133	Brilliant Blue FCF/FD&C Blue 1 (C.I. 42090)	Colouring	Synthetic blue colouring
140	Chlorophyll (C.I. 75810) *	Colouring	Natural olive to dark green colouring found in many plants. Solvents such as ethanol are used in the extraction of chlorophyll
141	Copper Complexes of Chlorophyll (C.I. 75810) *	Colouring	Olive-green colouring which is derived from the substitution of magnesium ion in chlorophyll with copper
142	Green S/Acid Brilliant Green BS/ Food green S/Lissamine green (C.I 44090)	Colouring	Synthetic green colouring
150	Caramel	Colouring	Dark brown colouring prepared by the controlled heat treatment of carbohydrates (e.g. glucose syrup, sucrose)
151	Brilliant Black BN (C.I. 28440)	Colouring	Synthetic black colouring
153	Carbon Black/Vegetable Carbon *	Colouring	Black colouring which may be prepared from animal charcoal or vegetable sources. The commercial source is mainly from plant material
154	Brown FK/Food Brown	Colouring	Synthetic brown colouring

Code	Chemical Name	Description	Remarks
155	Brown HT/Chocolate Brown HT (C.I. 20285)	Colouring	Synthetic brown colouring
160a	Alpha-/Beta-/Gamma-Carotene (C.I. 75130) *	Colouring	Orange-yellow colouring which is naturally found in many plants including in carrots. May be commercially synthesised in the laboratory
160b	Annatto/Bixin/Norbixin (C.I. 75120) *	Colouring	Yellow to peach colouring naturally found in the pericarp of the Annato tree. May be extracted by means of water/oil-soluble methods
160c	Capsanthin/Capsorubin/ Paprika extract *	Colouring	Red to orange colouring prepared by solvent extraction of the fruit pods and seeds of Capsicum annuum
160d	Lycopene (C.I. 75125) *	Colouring	Natural red colouring extracted from tomatoes
160e	Beta-apo-8'-carotenal/ Beta-8'-apocarotenal *	Colouring	Synthetic orange to yellowish-red colouring
160f	Ethyl ester of Beta-apo-8-carotenoic acid *	Colouring	Synthetic orange to yellow colouring
161a	Flavoxanthin *	Colouring	Yellow colouring which is a derivative of carotene (E160a). May contain other substances such as oils/ fats derived from the source material

Code	Chemical Name	Description	Remarks
161b	Lutein *	Colouring	Natural yellow to reddish colouring taken from plant extract. May also be obtained from the same source as chlorophyll (E140)
161c	Cryptoxanthin *	Colouring	Yellow colouring present in plants, orange rind, egg yolk and butter
161d	Rubixanthin *	Colouring	Yellow colouring present in rosehips
161e	Violaxanthin *	Colouring	Yellow colouring taken from plants e.g. yellow pansies
161f	Rhodoxanthin *	Colouring	Yellow colouring found in the seeds of the yew tree
161g	Canthaxanthin (C.I. 40850) *	Colouring	Orange colouring isolated from some mushrooms. May be commercially produced as part of the synthesis of carotene (E160a)
162	Beet Red/Betanin/Betanidin	Colouring	Deep purplish-red colouring extracted from beetroot
163	Anthocyanins *	Colouring	Red, blue colouring extracted from grape-skin extract and/or red cabbage by means of water, methanol or ethanol
170	Calcium Carbonate/Chalk (C.I. 77220) *	Inorganic colouring	May be extracted from naturally occurring white mineral or animal bones

Code	Chemical Name	Description	Remarks
171	Titanium Dioxide (C.I. 77891)	Inorganic colouring	White colouring prepared from naturally occurring mineral ilmenite
172	Iron Oxides/Red: 77491/Black: 77499 (Yellow: C.I. 77492)	Inorganic colouring	Yellow, red, orange, brown and black colouring from naturally occurring pigments or iron
173	Aluminium (C.I. 77000)	Inorganic colouring	Naturally occurring metallic colour from bauxite
174	Silver (C.I. 77820)	Inorganic colouring	Metallic colour from naturally occurring metal
175	Gold (C.I. 77480)	Inorganic colouring	Metallic colour from naturally occurring metal
180	Pigment Rubine/ Lithol Rubine BK (C.I. 15850)	Inorganic colouring	Synthetic red colouring
200	Sorbic Acid	Preservative	Naturally occurring in some fruits. May be synthetically manufactured from ketene
201	Sodium Sorbate	Preservative	Manufactured by neutralisation of sorbic acid (E200)
202	Potassium Sorbate	Preservative	Manufactured by neutralisation of sorbic acid (E200)
203	Calcium Sorbate	Preservative	Manufactured by neutralisation of sorbic acid (E200)
210	Benzoic Acid	Preservative	Naturally occurring in many edible berries, fruits and vegetables. May also be synthetically produced

Code	Chemical Name	Description	Remarks
211	Sodium Benzoate	Preservative	Sodium salt of benzoic acid (E210)
212	Potassium Benzoate	Preservative	Potassium salt of benzoic acid (E210)
213	Calcium Benzoate	Preservative	Calcium salt of benzoic acid (E210)
214	Ethyl 4-hydroxybenzoate/ Ethyl para-hydroxybenzoate	Preservative	Manufactured from benzoic acid (E210)
215	Ethyl 4-hydroxybenzoate, Sodium Salt /Sodium ethyl para-hydroxybenzoate	Preservative	Manufactured from benzoic acid (E210)
216	Propyl 4-hydroxybenzoate/ Propylparaben	Preservative	Manufactured from benzoic acid (E210)
217	Propyl 4-hydroxybenzoate, Sodium Salt	Preservative	Manufactured from benzoic acid (E210)
218	Methyl 4-hydroxybenzoate/ Methylparaben	Preservative	Manufactured from benzoic acid (E210)
219	Methyl 4-hydroxybenzoate, Sodium Salt	Preservative	Manufactured from benzoic acid (E210)

Code	Chemical Name	Description	Remarks
220	Sulphur Dioxide	Preservative	Manufactured chemically by the combustion of sulphur or gypsum
221	Sodium Sulphite	Preservative	Sodium salt of sulphurous acid
222	Sodium Hydrogen Sulphite/ Sodium Bisulphite	Preservative	Sodium salt of sulphurous acid
223	Sodium Metabisulphite	Preservative	Sodium salt of sulphurous acid
224	Potassium Metabisulphite/ Potassium	Pyrosulphite	Preservative Potassium salt of sulphurous acid
225	Potassium Sulphite	Preservative	Potassium salt of sulphurous acid
226	Calcium Sulphite	Preservative	Calcium salt of sulphurous acid
227	Calcium Hydrogen Sulphite/ Calcium Bisulphite	Preservative	Calcium salt of sulphurous acid
230	Biphenyl/Diphenyl	Preservative	Synthetically produced by action of heat on benzene
231	2-Hydroxybiphenyl	Preservative	Manufactured from phenyl ether or dibenzofuran
232	Sodium Biphenyl-2-yl-oxide	Preservative	Synthetically produced
233	2-(Thiazol-4-yl) Benzimidazole	Preservative	Chemically synthesised

Code	Chemical Name	Description	Remarks
234	Nisin*	Preservative	Produced by the growth of a bacterium called Streptococcus lactis.
235	Natamycin/Pimaricin*	Preservative	Produced by the growth of a bacterium called Strepmyces natalensis
236	Formic Acid	Preservative	Produced commercially by heating CO and NaOH under pressure and decomposing the resulting sodium formate with sulphuric acid
237	Sodium Formate	Preservative	Sodium salt of formic acid (E236)
238	Calcium Formate	Preservative	Calcium salt of formic acid (E236)
239	Hexamine	Preservative	Manufactured from formaldehdye and ammonia
249	Potassium Nitrite	Preservative	Potassium salt of nitrous acid
250	Sodium Nitrite	Preservative	Manufactured from sodium nitrate by bacterial or chemical actions
251	Sodium Nitrate	Preservative	Naturally occurring mineral
252	Potassium Nitrate/Saltpetre*	Preservative	Naturally occurring mineral. May also be artificially produced from vegetable material and waste animal

Code	Chemical Name	Description	Remarks
260	Acetic Acid*	Food Acid	Commercially manufactured by the action of CH_3OH and CO. The CH_3COOH in vinegar may be produced by the action of bacterium Acetobacter on alcohol
261	Potassium Acetate	Food Acid	Potassium salt of acetic acid (E260)
262	Sodium Acetate	Food Acid	Sodium salt of acetic acid (E260)
263	Calcium Acetate	Food Acid	Calcium salt of acetic acid (E260)
264	Ammonium Acetate	Food Acid	Ammonium salt of acetic acid (E260)
270	Lactic Acid*	Food Acid	Commercially produced by heat treatment of carbo[illegible], such as whey, and fermented by bacteria such as Bacillus acidilacti, Lactobacillus delbueckii or L. bulgaricus
280	Propionic Acid*	Preservative	May be commercially derived from natural gas or from wood pulp waste liquor by the fermentation activity of Propionibacteria
281	Sodium Propionate*	Preservative	Sodium salt of propionic acid (E280)
282	Calcium Propionate*	Preservative	Calcium salt of propionic acid (E280)
283	Potassium Propionate*	Preservative	Potassium salt of propionic acid (E280)

Code	Chemical Name	Description	Remarks
290	Carbon Dioxide	Propellant	Naturally occuring. May also be produced by way of fermentation or acid-carbonate reaction
296	Malic Acid (DL- or L-)	Food Acid	Commercially synthesised by means of heating malic with sulphuric acid
297	Fumaric Acid*	Food Acid	Commercially prepared by glucose fermentation using fungi such as Rhizopus nigricans
300	L-Ascorbic Acid/Vitamin C*	Antioxidant/ Vitamin	Occurs naturally in many fruits and vegetables. May be synthesised from the hydrogenation of glucose to sorbitol and its eventual conversion to ascorbic acid. May also be biologically synthesised through means of fermentation
301	Sodium Ascorbate	Antioxidant/ Vitamin	Synthetic sodium salt of ascorbic acid (E300)
302	Calcium Ascorbate	Antioxidant/ Vitamin	Synthetic calcium salt of ascorbic acid (E300)
303	Potassium Ascorbate	Antioxidant/ Vitamin	Synthetic potassium salt of ascorbic acid (E300)

Code	Chemical Name	Description	Remarks
304	Ascorbyl Palmitate*	Antioxidant/ Vitamin	Ascorbic acid ester comprising ascorbic acid and palmitic acid
306	Tocopherol Concentrate, Mixed/Vitamin E	Antioxidant/ Vitamin	Extracts from soya bean oil, rice germ, wheat germ, maize and green leaves
307	Synthetic Alpha-Tocopherol	Antioxidant/ Vitamin	Manufactured by chemical synthesis
308	Synthetic Gamma-Tocopherol	Antioxidant/ Vitamin	Manufactured by chemical synthesis
309	Synthetic Delta-Tocopherol	Antioxidant/ Vitamin	Manufactured by chemical synthesis
310	Propyl Gallate	Antioxidant	Manufactured from gallic acid found in the tannins of nut galls. May also be produced from the hydrolysis of tannase, which may occur in spent fungal broth
311	Octyl Gallate	Antioxidant	Manufactured from gallic acid found in the tannins of nut galls. May also be produced from the hydrolysis of tannase, which may occur in spent fungal broth

Code	Chemical Name	Description	Remarks
312	Dodecyl Gallate	Antioxidant	Manufactured from gallic acid found in the tannins of nut galls. May also be produced from the hydrolysis of tannase, which may occur in spent fungal broth
317	Erythorbic Acid/ Iso-ascorbic Acid*	Antioxidant	Commercially produced from sucrose by fermentation with Penicillium sp.
318	Sodium Erythorbate/ Sodium Iso-Ascorbate*	Antioxidant	Sodium salt of erythorbic acid (E317)
319	tert-Butylhydroquinone/TBHQ	Antioxidant	Derived from petroleum
320	Butylated Hydroxyanisole (BHA)*	Antioxidant	Commercially prepared from p-methoxyphenol and isobutene
321	Butylated Hydroxytoluene (BHT)*	Antioxidant	Prepared synthetically from p-cresol and isobutylene
322	Lecithins*	Emulsifier/ Antioxidant	Obtained from animal or vegetable materials through physical procedures. Most lecithin are commercially obtained from soya beans
325	Sodium Lactate*	Food Acid	Sodium salt of lactic acid (E270)
326	Potassium Lactate*	Food Acid	Potassium salt of lactic acid (E270)

Code	Chemical Name	Description	Remarks
327	Calcium Lactate*	Food Acid	Calcium salt of lactic acid (E270)
328	Ammonium Lactate*	Food Acid	Ammonium salt of lactic acid (E270)
329	Magnesium Lactate*	Food Acid	Magnesium salt of lactic acid (E270)
330	Citric Acid*	Food Acid	Commercially prepared by the fermentation of molasses with fungal strains of Aspergillus niger. May also be isolated from pineapple by-products and low-grade lemons
331	Sodium Citrates*	Food Acid	Sodium salt of citric acid (E330)
332	Potassium Citrates*	Food Acid	Potassium salt of citric acid (E330)
333	Calcium Citrates*	Food Acid	Calcium salt of citric acid (E330)
334	Tartaric Acid*	Food Acid	Most commercially available tartaric acid is manufactured as a by-product of the wide industry. May also be extracted from tamarind pulp
335	Sodium Tartrate*	Food Acid	Sodium salt of tartaric acid (E334)
336	Potassium Tartrate/ Potassium Hydrogen Tartrate/ Cream of Tartar*	Food Acid	By-product of the wine industry

Code	Chemical Name	Description	Remarks
337	Potassium Sodium Tartrate*	Food Acid	Derivative of tartaric acid (E334)
338	Phosphoric Acid/ Orthophosphoric Acid	Miscellaneous	Manufactured from phosphate ore
339	Sodium Phosphates	Mineral Salt	Sodium salt of phosphoric acid (E338)
340	Potassium Phosphates	Mineral Salt	Potassium salt of phosphoric acid (E338)
341	Calcium Phosphates	Mineral Salt	Calcium salt of phosphoric acid (E338)
343	Magnesium Phosphates	Mineral Salt	Naturally occurring mineral
350	Sodium Malate/ Sodium Hydrogen Malate	Food Acid	Sodium salt of malic acid (E296)
351	Potassium Malate	Food Acid	Potassium salt of malic acid (E296)
352	Calcium Malate/Calcium Hydrogen Malate	Food Acid	Calcium salt of malic acid (E296)
353	Metatartaric Acid*	Sequestrant	Prepared from tartaric acid (E334)
355	Adipic Acid/Hexanedioic Acid	Buffer	Commercially produced by oxidising cyclohexanol with concentrated nitric acid
357	Potassium Adipate	Buffer	Potassium salt of adipic acid (E355)

Code	Chemical Name	Description	Remarks
363	Succinic Acid*	Buffer/ Food Acid*	Commercially prepared from acetic acid (E260)
365	Sodium Fumarate*	Food Acid	Sodium salt of fumaric acid (E297)
366	Potassium Fumarate*	Food Acid	Potassium salt of fumaric acid (E297)
367	Calcium Fumarate*	Food Acid	Calcium salt of fumaric acid (E297)
370	1,4-Heptonolactone	Sequestrant	Prepared from hydroxycarboxylic acid
375	Niacin/Nicotinic Acid/ Nicotinamide	Vitamin	Commercially prepared by the oxidation of nicotine with concentrated nitric acid
380	Triammonium Citrate*	Buffer	Ammonium salt of citric acid (E330)
381	Ammonium Ferric Citrate*	Dietary Supplement	Prepared from citric acid (E330)
385	Calcium Disodium EDTA	Chelating Agent	Synthetically prepared
400	Alginic Acid	Vegetable Gum	Extracted from brown seaweeds such as the species of Laminaria, Macrocystis and Ascophyllum
401	Sodium Alginate	Vegetable Gum	Sodium salt of alginic acid (E400)
402	Potassium Alginate	Vegetable Gum	Potassium salt of alginic acid (E400)

Code	Chemical Name	Description	Remarks
403	Ammonium Alginate	Vegetable Gum	Ammonium salt of alginic acid (E400)
404	Calcium Alginate	Vegetable Gum	Calcium salt of alginic acid (E400)
405	Propane-1,2-Diol Alginate/ Propylene Glycol Alginate/ Alginate Ester	Vegetable Gum	Propylene glycol ester of alginic acid (E400)
406	Agar/Agar-Agar/ Japanese Isinglass	Vegetable Gum	Extracted from red seaweeds such as the Gelidium amansii. May also be taken from members of the related red algae Rhodophyceae
407	Carrageenan/Irish Moss	Vegetable Gum	Occurs naturally in red seaweeds belonging to the Gigartinaceae, Solieriaceae, Hypnaceae and Furcellariaceae families
410	Locust Bean Gum/ Carob Bean Gum	Vegetable Gum	Taken from the Locust or Carob tree (Ceratonia siliqua), which is an evergreen tree belonging to the Leguminoseae or pea family
412	Guar Gum	Vegetable Gum	Extracted from the seeds of Cyamopsis tetragonolobus, or C. psoraloides, a member of the pea family
413	Tragacanth/Gum Tragacanth	Vegetable Gum	Extracted from the trunk and branches of Astragalus gummifier and other species of the pea family

Code	Chemical Name	Description	Remarks
414	Acacia/Gum Arabic	Vegetable Gum	Occurs naturally in the stems and branches of Acacia senegal and members of the pea family
415	Xanthan Gum/Corn Sugar Gum*	Vegetable Gum	Produced by the fermentation of carbohydrate using a bacterium known as Xanthomonas campestris
416	Karaya Gum/Sterculia Gum	Vegetable Gum	Occurs naturally in the trunk and stem of the tree Sterculia urens
420	Sorbitol/Sorbitol Syrup*	Humectant	Commercially produced from glucose by hydrogenation or electrolytic reduction
421	Mannitol/Manna Sugar	Humectant	Prepared from seaweed or manna, the dried exudate of Fraxinus ornus. May be commercially prepared by the hydrogenation of invert sugar, monosaccharides and sucrose
422	Glycerol/Glycerin*	Humectant	Industrial by-product in the manufacture of soaps, candles and fatty acids from oils and fats. May also be synthesised from propylene or by the fermentation of sugars
430	Polyoxyethylene (8) Stearate*	Emulsifier	Synthesised using stearic acid (E570)
431	Polyoxyethylene (40) Stearate*	Emulsifier	Synthesised using stearic acid (E570)

Code	Chemical Name	Description	Remarks
432	Polyoxyethylene (20) Sorbitan Monolaurate/ Polysorbate 20/Tween 20*	Emulsifier	Lauric ester of sorbitol and sorbitol anhydride
433	Polyoxyethylene (20) Sorbitan Mono-Oleate /Polysorbate 80/Tween 80*	Emulsifier	Oleic ester of sorbitol and sorbitol anhydride
434	Polyoxyethylene (20) Sorbitan Monopalmitate /Polysorbate 40/Tween 40*	Emulsifier	Palmitate ester of sorbitol and sorbitol anhydride
435	Polyoxyethylene (20) Sorbitan Monostearate /Polysorbate 60/Tween 60*	Emulsifier	Stearic acid ester of sorbitol and sorbitol anhydride
436	Polyoxyethylene (20) Sorbitan Tristearate /Polysorbate 65/Tween 65 *	Emulsifier	Stearic acid ester of sorbitol and sorbitol anhydride
440a	Pectin	Stabiliser/ Thickener	Apple residues and orange pith are commercial sources of pectin

Code	Chemical Name	Description	Remarks
440b	Amidated Pectin	Emulsifier/ Stabiliser	Derived from the treatment of pectin (E440a) with ammonia
441	Gelatine*	Emulsifier/ Stabiliser	Obtained by boiling animal skin (usually cattle or pig's), ligaments, bones, or any tissue that contains collagen
442	Ammonium Phosphatides/ Emulsifier YN	Emulsifier/ Stabiliser	Prepared synthetically
450	Na and K Metaphosphates, Polyphosphates and Pyrophosphates	Mineral Salt	Sodium and potassium salt of phosphoric acid (E338)
460	Microcrystalline Cellulose/ Powdered Cellulose	Anticaking Agent	Prepared from the cellulose component of plant cell wall
461	Methylcellulose/Cologel/ Methocel	Vegetable Gum	Prepared from plant cellulose
463	Hydroxypropylcellulose	Vegetable Gum	Prepared from plant cellulose
464	Hydroxypropyl-Methylcellulose	Vegetable Gum	Prepared from plant cellulose
465	Ethylmethylcellulose	Vegetable Gum	Prepared from plant cellulose

Code	Chemical Name	Description	Remarks
466	Sodium Carboxymethylcellulose/ CMC	Vegetable Gum	Prepared from plant cellulose
469	Sodium Caseinate	Emulsifier/ Stabiliser	Derived from the protein of cow's milk
470	Na, K and Ca Salts of Fatty Acids*	Emulsifier/ Stabiliser	Prepared from fatty acids
471	Mono-and Diglycerides of Fatty Acids*	Emulsifier/ Stabiliser	Commercially prepared from glycerin (E422) and fatty acids
472	Various Esters of Glycerol*	Emulsifier/ Stabiliser	Prepared from esters of glycerol and fatty acids
473	Sucrose Esters of Fatty Acids*	Emulsifier/ Stabiliser	Prepared from esters of glycerol and sucrose
474	Sucroglycerides*	Emulsifier/ Stabiliser	Prepared by the action of sucrose on natural triglycerides (from lard, tallow, palm oil, etc)
475	Polyglycerol Esters of Fatty Acids*	Emulsifier/ Stabiliser	Prepared in the laboratory

Code	Chemical Name	Description	Remarks
476	Polyglycerol Esters of Polycendensed Fatty Acids	Emulsifier/ Stabiliser	Prepared from castor oil and glycerol esters of Castor Oil/Polyglycerol Polyricinoleate*
477	Propane-1,2-Diol Esters of Fatty Acids*	Emulsifier/ Stabiliser	Prepared from propylene glycol
481	Sodium Stearoyl-2-Lactylate*	Emulsifier/ Stabiliser	Prepared from lactic acid (E270)
482	Calcium Stearoyl-2-Lactylate*	Emulsifier/ Stabiliser	Prepared from lactic acid (E270)
483	Stearyl Tartrate*	Emulsifier/ Stabiliser	Prepared from tartaric acid (E334)
491	Sorbitan Monostearate*	Emulsifier/ Stabiliser	Prepared synthetically from stearic acid (E570) and sorbitol (E420)
492	Sorbitan Tristearate/Span 65*	Emulsifier/ Stabiliser	Prepared synthetically from stearic acid (E570) and sorbitol (E420)
493	Sorbitan Monolaurate/Span 20*	Emulsifier/ Stabiliser	Prepared synthetically from sorbitol (E420) and lauric acid

Code	Chemical Name	Description	Remarks
494	Sorbitan Monooleate/Span 80*	Emulsifier/ Stabiliser	Prepared synthetically from sorbitol (E420) and oleic acid
495	Sorbitan Monopalmitate/ Span 40*	Emulsifier/ Stabiliser	Prepared synthetically from sorbitol (E420) and palmitic acid
500	Na Carbonate/Na Bicarbonate/ Baking Soda	Mineral Salt	Synthetically prepared. May also be manufactured by the Solvay process or electrolytically from sea water
501	K Carbonate/ K Hydrogen Carbonate	Mineral Salt	Prepared by saturating a concentrated solution of potassium carbonate with carbon dioxide
503	NH4 Bicarbonate/ NH4 Hydrogen Carbonate	Mineral Salt	Prepared by passing excess carbon dioxide through concentrated ammonia water
504	Magnesium Carbonate	Mineral Salt	May be prepared by mixing boiling concentrated solutions of magnesium sulphate and sodium carbonate
507	Hydrochloric Acid	Food Acid	Industrially produced by the reaction of sodium chloride and sulphuric acid
508	Potassium Chloride	Salt Substitute	Naturally occurs as a saline residue associated with rock salt

Code	Chemical Name	Description	Remarks
509	Calcium Chloride	Mineral Salt	Obtained as a by-product of the Solvay process and is also a product from natural salt brines
510	Ammonium Chloride	Flour Treatment Agent	Synthetically prepared
511	Magnesium Chloride	Firming Agent	Prepared from magnesium ammonium chloride hexahydrate, in the presence of hydrochloric acid
513	Sulphuric Acid	Food Acid	Commercially prepared by the 'contact' or 'chamber' process
514	Sodium Sulphate	Diluent	Naturally occurring
515	Potassium Sulphate	Salt Substitute	Naturally occurring
516	Calcium Sulphate	Flour Treatment Agent	Naturally occurring
518	Magnesium Sulphate	Dietary Suppl./ Firming Agent	Naturally occurring
519	Cupric Sulphate/ Copper Sulphate	Preservative/ Colour Fixative	Industrially prepared by spraying hot dilute sulphuric acid on to scrap copper in a lead-lined tower

Code	Chemical Name	Description	Remarks
524	Sodium Hydroxide	Neutralising Agent	Prepared by electrolysis from brine, or precipitated from sodium carbonate and lime solution
525	Potassium Hydroxide	Oxidising Agent	Industrially prepared by electrolysis of potassium chloride
526	Calcium Hydroxide	Neutralising Agent	Prepared by the hydration of lime
527	Ammonium Hydroxide	Alkali	Prepared from ammonia gas
528	Magnesium Hydroxide	Alkali	Commercially prepared from magnesite ores
529	Calcium Oxide	Alkali	Prepared from limestone
530	Magnesium Oxide	Alkali	Commercially prepared from magnesite ores
535	Sodium Ferrocyanide	Anticaking Agent	Synthetically produced
536	Potassium Ferrocyanide	Anticaking Agent	Commercially prepared as a by-product in the purification of coal gas
540	Dicalcium Diphosphate	Buffer	Naturally occurring mineral. May also be synthetically produced
541	Sodium Aluminium Phosphate	Aerator/ Emulsifying Salt	Prepared from phosphoric acid (E338)

Code	Chemical Name	Description	Remarks
542	Edible Bone Phosphate/ Bone Meal*	Anticaking Agent	Extract from animal bones
544	Calcium Polyphosphates*	Emulsifying Salt	Calcium salts of polyphosphoric acid
545	Ammonium Polyphosphates	Emulsifying Salt	Ammonium salts of polyphosphoric acid
551	Silicon Dioxide/Silica Salt	Anticaking Agent	Rock-forming mineral and sand which is composed mainly of quartz or flint
552	Calcium Silicate	Anticaking Agent	Commercially prepared from lime and diatomaceous earth
553a	Magnesium Silicate/ Magnesium Trisilicate	Anticaking Agent	Synthetic compound of magnesium oxide and silicon dioxide. May also be prepared from sodium silicate and magnesium sulphate
553b	Talc	Anticaking Agent	Naturally occurring mineral
554	Aluminium Na Silicate/ Na Aluminosilicate	Anticaking Agent	Prepared synthetically from quartz and gibbsite
556	Aluminium Calcium Silicate	Anticaking Agent	Naturally occurring mineral
558	Bentonite	Anticaking Agent	Naturally occurring
559	Kaolin	Anticaking Agent	Naturally occurring

Code	Chemical Name	Description	Remarks
570	Stearic Acid*	Anticaking Agent	Naturally occurring fatty acid found in all animal fats and vegetable oils. May be prepared synthetically for commercial use
572	Magnesium Stearate*	Anticaking Agent	Magnesium salt of stearic acid (E570)
575	Glucono Delta-Lactone*	Sequestrant	Prepared by the oxidation of glucose
576	Sodium Gluconate	Sequestrant	Sodium salt of gluconic acid
577	Potassium Gluconate	Sequestrant	Potassium salt of gluconic acid
578	Calcium Gluconate	Sequestrant	Calcium salt of gluconic acid
579	Ferrous Gluconate	Colouring/ Flavouring	Prepared from barium gluconate and ferrous sulphate
620	L-Glutamic Acid*	Flavour Enhancer	Commercially prepared by the fermentation of carbohydrate by a bacterium e.g. Micrococcus glutamicus
621	Monosodium Glutamate/MSG*	Flavour Enhancer	Sodium salt of glutamic acid (E620)
622	Monopotassium Glutamate*	Flavour Enhancer	Potassium salt of glutamic acid (E620)
623	Calcium Glutamate*	Flavour Enhancer	Calcium salt of glutamic acid (E620)

Code	Chemical Name	Description	Remarks
627	Disodium Guanylate*	Flavour Enhancer	Sodium salt of guanylic acid, a widely occurring nucleotide found in yeast extract and sardines. May be synthetically prepared from commercial use
631	Disodium Inosinate*	Flavour Enhancer	Sodium salt of inosinic acid, found in meat extract and sardines
635	Sodium 5'-Ribonucleotide*	Flavour Enhancer	Mixture of disodium guanylate (E627) and disodium inosinate (E631)
636	Maltol	Flavour Enhancer	Occurs naturally in the bark of larch trees, pine needles and roasted malt. May also be obtained by the alkaline hydrolysis of streptomycin salt
637	Ethyl Maltol	Flavour Enhancer	Prepared from maltol (E636)
900	Dimethylpolysiloxane/ Dimethicone	Antifoaming Agent	A mixture of liquid dimethylpolysiloxane and silicon gel or silicon dioxide
901	Beeswax *	Glazing Agent	Naturally occurring from bee honeycomb. White beewax is bleached and purified
903	Carnauba Wax	Glazing Agent	Obtained from the surface of leaves of Copernicia cerifera, a Brazilian wax palm

Code	Chemical Name	Description	Remarks
904	Shellac *	Glazing Agent	Obtained from the resin produced by lac insect (Laccifer lacca)
905	Mineral Oil/Petrolatum	Glazing Agent	Derived from petroleum
907	Refined Microcrystalline Wax	Glazing Agent	Derived from petroleum
920	L-Cysteine Hydrochloride*	Flour Treatment	Manufactured from animal hair and chicken feathers Agent
924	Potassium Bromate	Flour Treatment Agent	Synthetically produced
925	Chlorine	Preservative/ Bleaching Agent	Commercially produced by electrolysis
926	Chlorine Dioxide	Bleaching/ Improving Agent	Synthetically prepared
927	Azodicarbonamide/ Azoformamide	Improving Agent	Synthetically prepared
928	Benzoyl Peroxide/ Dibenzoyl Peroxide	Bleaching Agent	Synthetically prepared
931	Nitrogen	Propellant	Industrially produced by the reduction of ammonia or by the fractional distillation of liquid air

Code	Chemical Name	Description	Remarks
932	Nitrous Oxide	Propellant	Industrially produced by the thermal decomposition of ammonium nitrate
950	Acesulphame Potassium/ Sunett	Artificial Sweetener	Potassium salt of 6-methyl-1,2,3-oxathiazin-4 (3H)-1,2,2-dioxide
951	Aspartame/Nutrasweet*	Artificial Sweetener	Commercially produced by combining two amino acids together, namely L-phenylalanine and L-aspartic acid
952	Cyclamic and its Calcium and Sodium Salts	Artificial Sweetener	Manufactured by many different methods
954	Saccharin and its Calcium and Sodium Salts	Artificial Sweetener	Manufactured by many different methods
957	Thaumatin	Artificial Sweetener	Derived from an African plant called Thaumococcus danielli
965	Hydrogenated Glucose Syrup*	Humectant	Derived from starches, which originate from many different sources, and broken down by enzymes and water to form glucose, oligosaccharides, followed by maltitol and sorbitol (E420)
967	Xylitol	Sweetener	Commercially produced as a waste product of the pulp industry

Code	Chemical Name	Description	Remarks
1200	Polydextrose*	Miscellaneous	Manufactured from glucose, citric acid and sorbitol
1201	Polyvinylpyrrolidone	Miscellaneous	Commercially produced from acetylene, hydrogen, formaldehyde and ammonia
1202	Polyvinyl Polypyrrolidone	Miscellaneous	It is the insolule form of polyvinylpyrrolidone (E1201)
1400	Dextrin	Thickener	May be produced by the dry heating of unmodified starch or in the presence of acids and buffers. Starches used are mainly from corn (maize) and tapioca
1403	Bleached Starch	Thickener	May be obtained by treating native starch with various chemicals including hydrogen peroxide, sodium chlorite or sulphur dioxide
1404	Oxidised Starch	Thickener	Produced by treating native starch with sodium hypochlorite
1405	Enzyme-treated Starch*	Thickener	Produced by subjecting corn starch to acid-enzyme treatment to yield glucose, maltose and higher oligosaccharides
1410	Monostarch Phosphate	Thickener	Produced by the esterification of native starch with orthophosphoric acid, Na or K orthophosphate, or sodium tripolyphosphate

Code	Chemical Name	Description	Remarks
1412	Distarch Phosphate	Thickener	Produced by the esterification of native starch with sodium trimetaphosphate or phosphorus oxychloride
1413	Phosphated Distarch Phosphate	Thickener	Produced by the esterification of native starch with phosphate, and dually stabilised with a 'monosubstituent group' of phosphate
1414	Acetylated Distarch Phosphate	Thickener	Produced by the esterification of native starch with sodium trimetaphosphate or phosphorus oxychloride, and stabilised with a 'monosubstituent group' of acetate
1420	Starch Acetate Esterified with Acetic Anhydride	Thickener	Produced by the esterification of native starch with a mixed anhydride of adipic and acetic anhydride, and stabilised with a 'monosubstituent group' of acetate
1421	Starch Acetate Esterified with Vinyl Acetate	Thickener	Produced by the esterification of native starch with monosubstituent groups of vinyl acetate
1422	Acetylated Distarch Adipate	Thickener	Produced by the esterification of native starch with a mixed anhydride of adipic and acetic anhydride
1440	Hydroxypropyl Starch	Thickener	Produced by treating native starch with the hydroxypropyl group

Code	Chemical Name	Description	Remarks
1442	Hydroxypropyl Distarch Phosphate	Thickener	Produced by the esterification of native starch with phosphate, and stabilised with a monosubstituent hydroxyl group
1450	Starch, Sodium Octenylsuccinate	Thickener	Produced by treating native starch with an octenylsuccinate half ester monosubstituent group
1505	Triethyl Citrate/ Ethyl Citrate	Miscellaneous	Bitter oily liquid which is soluble in water and can be mixed with alcohol
1510	Ethyl Alcohol/Ethanol*	Miscellaneous	Produced by the fermentation of carbohydrates. May also be obtained from ethylene, acetylene or liquors from waste sulphites. Other manufacturing process includes hydrolysis of ethyl sulphate or by the oxidation of methane
1518	Triacetin/Glycerol Triacetate*	Miscellaneous	Produced by the acetylation of glycerol
1520	Propylene Glycol*	Miscellaneous	Commercially produced from propylene, or by heating glycerol with sodium hydroxide, or by reacting propylene oxide with water

*Doubtful (either by way of its source or manufacturing process

Appendix-X

Fundamental Physical Constants

Constant	Symbol	Value	
electron mass	m_e	9.10939×10^{-31}	kg
proton mass	m_p	1.67262×10^{-27}	kg
neutron mass	m_n	1.67493×10^{-27}	kg
atomic mass unit	u	1.66054×10^{-27}	kg
elementary charge	e	1.60218×10^{-19}	C
Avogadro constant	N_A	6.02214×10^{23}	mol^{-1}
molar gas constant	R	8.31451	$J\ mol^{-1}\ K^{-1}$
standard atmosphere	atm	1.01325×10^{5}	Pa
molar volume	V_m	22.41410	L mol-1
speed of light	c	2.99792×10^{8}	$m\ s^{-1}$
Plank constant	h	6.62608×10^{-34}	J s
Faraday constant	F	9.64853×10^{4}	$C\ mol^{-1}$
Boltzmann constant	K_B	1.38066×10^{-23}	$J\ K^{-1}$
conductance quantum	G_0	7.748×10^{-5}	S

Constant	Symbol	Value	
electric constant	E_0	8.854×10^{-12}	F m-1
fine-structure constant	α	7.297×10^{-3}	
inverse finestructure constant	A^{-1}	137.035999	
proton-electron mass ratio	m_p/m_e	1836.153	
Rydberg constant	R	10 973 731.57	m-1
Newtonian constant of gravitation	G	6.67×10^{-11}	$m^3\,kg^{-1}\,s^{-2}$

Appendix-XI

Generation Times: Selected Microorganisms

Microorganism	Temperature (°F)	Generation time (Hours)
Bacteria		
Beneckea natriegens	98.6	0.16
Escherichia coli	104	0.35
Bacillus subtilis	104	0.43
Staphylococcus aureus	98.6	0.47
Pseudomonas aeruginosa	98.6	0.58
Clostridium botulinum	98.6	0.58
Rhodospirillum rubrum	77	4.6-5.3
Anabaena cylindrica	77.1	10.6
Mycobacterium tuberculosis	98.6	~12
Treponema pallidum	98.6	33

Microorganism	**Temperature (°F)**	**Generation time (Hours)**
Algae		
Scenedesmus quadricauda	77.2	5.9
Chlorella pyrenoidosa	77	7.75
Asterionella formosa	68	9.6
Euglena gracilis	77	10.9
Ceratium tripos	68	82.8
Protozoa		
Tetrahymena geleii	75.2	2.2-4.2
Leishmania donovani	78.8	10-12
Paramecium caudatum	78.8	10.4
Acanthamoeba castellanii	86.3	11-12
Giardia lamblia	98.6	18
Fungi		
Saccharomyces cerevisiae	86.1	2
Monilinia fructicola	77	30

Appendix-XII

Genetic Codes: RNA Form

Second position

	U	C	A	G	
U	UUU Phe	UCU Ser	UAU Tyr	UGU Cys	**U**
	UUC Phe	UCC Ser	UAC Tyr	UGC Cys	**C**
	UUA Leu	UCA Ser	UAA STOP	UGA STOP	**A**
	UUG Leu	UCG Ser	UAG STOP	UGG Trp	**G**
C	CUU Leu	CCU Pro	CAU His	CGU Arg	**U**
	CUC Leu	CCC Pro	CAC His	CGC Arg	**C**
	CUA Leu	CCA Pro	CAA Gln	CGA Arg	**A**
	CUG Leu	CCG Pro	CAG Gln	CGG Arg	**G**
A	AUU Lle	ACU Thr	AAU Asn	AGU Ser	**U**
	AUC Lle	*ACC Thr*	*AAC Asn*	*AGC Ser*	***C***
	AUA Lle	ACA Thr	AAA Lys	AGA Arg	**A**
	AUG Met	ACG Thr	AAG Lys	AGG Arg	**G**

G	GUU Val	GCU Ala	GAU Asp	GGU Gly	**U**
	GUC Val	GCC Ala	GAC Asp	GGC Gly	**C**
	GUA Val	GCA Ala	GAA Glu	GGA Gly	**A**
	GUG Val	GCG Ala	GAG Glu	GGG Gly	**G**

A-Adenine; T-Thymine; G-Guanine; C-Cytosine

Ala-Alanine; Arg- Arginine; Asn- Asparagine; Asp-Aspartic acid; Cys-Cysteine; Gln-Glutamine; Glu-Glutamic acid; Gly-Glycine; His-Histidine; Ile-Isoleucine; Leu-Leucine; Lys-Lysine; Met-Methionine; Phe-Phenylalanin, Pro-Proline; Ser- Serine; Thr-Threonine; Trp-Trptophan; Tyr-Tyrosine; Val-Valine

Appendix-XIII

Genetic Material: Common Viruses

SS/DS	Structure	example/s
DNA		
Single-Stranded	Linear single strand	Parvoviruses
	Circular single strand	φX174, M13, fd phages
Double-Stranded	Linear double strand	Herpesviruses (herpes simplex viruses, cytomegalovirus, Epstein-Barr virus), adenoviruses, T coliphages, lambda phage, and other bacteriophages
	Linear double strand with single chain breaks	T5 coliphage
	Double strand with cross-linked ends	Vaccinia, smallpox

SS/DS	Structure	example/s
	Closed circular double strand	Polyomaviruses (SV-40), papillomaviruses, PM2 phage, cauliflower mosaic
RNA		
Single-Stranded	Linear, single stranded, positive strand	Picornaviruses (polio, rhinoviruses), togaviruses, RNA bacteriophages, TMV, and most plant viruses
	Linear, single stranded, negative strand	Rhabdoviruses (rabies), paramyxoviruses (mumps, measles)
	Linear, single stranded, segmented, positive strand	Brome mosaic virus (individual segments in separate virions)
	Linear, single stranded, segmented, diploid (two identical single strands), positive strand	Retroviruses (Rous sarcoma virus, human immunodeficiency virus)
	Linear, single stranded, segmented, negative strand	Paramyxoviruses, orthomyxoviruses (influenza)
Double-Stranded	Linear, double stranded, segmented	Reoviruses, wound-tumor virus of plants, cytoplasmic polyhedrosis virus of insects, phage φ6, many mycoviruses

Appendix-XIV

Human Immunoglobulins

Immunoglobulin Classes

Property	IgG[a]	IgM	IgA[b]	IgD	IgE
Heavy chain	γ_1	1	α_1	δ	ε
Mean serum concentration (mg/ml)	9	1.5	3.0	0.03	0.00005
Percent of total antibody	80-85	5-10	5-15	<1	0.002-0.05
Valency	2	5(10)	(2- 4)	2	2
Mass of heavy chain (kDa)	51	65	56	70	72
Mass of entire molecule (kDa)	146	970	160[c]	184	188
Placental transfer	+	-	-	-	-
Half-life in serum (days)[d]	23	5	6	3	2

Property	IgG[a]	IgM	IgA[b]	IgD	IgE
Complement activation	++	+++	-	-	-
Classical pathway					
Alternative pathway	-	-	+	-	-
Induces mast cell degranulation	-	-	-	-	+
% carbohydrate	3	7-10	7	12	11
Major characteristics	Most abundant Ig in body fluids; neutralizes toxins, opsonizes bacteria, activates complement, maternal antibody	First to appear after antigen stimulation; very effective agglutinator; expressed as membrane-bound antibody on B Cells	Secretory antibody; protects external surfaces	Present on B-cell surface; B-cell recognition of antigen	Anaphylactic-mediating antibody: resistance to helminths

[a]Properties of IgG subclass 1; [b]Properties of IgA subclass 1; [c]sIgA = 360 - 400 kDa; [d]Time required for half of the antibodies to disappear.

Appendix-XV

Inhibition Zone Diameter: Chemotherapeutic Drugs

Chemotherapeutic Drug	*Zone Diameter (Nearest mm)*			
	Disk Content	Resistant	Intermediate	Susceptible
Carbenicillin (with Proteus spp. and E. coli)	100 µg	≥17	18-22	≥23
Carbenicillin (with Pseudomonas aeruginosa)	100 µg	≥13	14-16	≥17
Ceftriaxone	30 µg	≥13	14-20	≥21
Chloramphenicol	30 µg	≥12	13-17	≥18
Erythromycin	15 µg	≥13	14-17	≥18
Penicillin G (with staphylococci)	10 U[a]	≥20	21-28	≥29

Chemotherapeutic Drug	Disk Content	Resistant	Intermediate	Susceptible
Penicillin G (with other microorganisms)	10 U	≥11	12-21	≥22
Streptomycin	10 µg	≥11	12-14	≥15
Sulfonamides	250 or 300 µg	≥12	13-16	≥17
Tetracycline	30µg	≥14	15-18	≥19

[a]One milligram of penicillin G sodium = 1,600 units.

Appendix-XVI

Ion Names & Formulae

Monatomic		Polatomic	
Ag^{+}	silver ion	BO_3^{3-}	borate ion
Al^{3+}	aluminum ion	$C_2H_3O_2^{-}$	acetate ion
Au^{+} / Au^{2+}	gold ion	ClO^{-}	hypochlorite ion
Be^{2+}	beryllium ion	ClO_2^{-}	chlorite ion
Ca^{2+}	calcium ion	ClO_3^{-}	chlorate ion
Co^{2+} / Co^{3+}	cobalt ion	ClO_4^{-}	perchlorate ion
Cr^{2+} / Cr^{3+}	chromium ion	CN^{-}	cyanide ion
Cu^{+} / Cu^{2+}	copper ion	CO_3^{2-}	carbonate ion
Fe^{2+} / Fe^{3+}	iron ion	$C_2O_4^{2-}$	oxalate ion
K^{+}	potassium ion	CrO_4^{2-}	chromate ion
Li^{+}	lithium ion	$Cr_2O_7^{2-}$	dichromate ion
Mg^{2+}	magnesium ion	HCO_3^{-}	hydrogen carbonate/ bicarbonate
Na^{+}	sodium ion	H_3O^{+}	hydroniumion

Monatomic		**Polatomic**	
Zn^{2+}	zinc ion	HPO_4^{2-}	hydrogen phosphate ion
		$H_2PO_4^-$	dihydrogen phosphate ion
		HSO_3^-	hydrogen sulphite/ bisulphite
		HSO_4^-	hydrogen sulphate/ bisulphate
		MnO_4^-	permanganate ion
		N_3^-	azide ion
		NH_4^+	ammonium ion
		NO_2^-	nitrite ion
		NO_3^-	nitrate ion
		O_2^{2-}	peroxide ion
		OCN^-	cyanate ion
		OH^-	hydroxide ion
		PO_3^{3-}	phosphite ion
		PO_4^{3-}	phosphate ion
		SCN^-	thiocyanate ion
		SO_3^{2-}	sulphite ion
		SO_4^{2-}	sulphate ion
		$S_2O_3^{2-}$	thiosulphateion

Appendix-XVII

ISO Codes for important Food/ Food Product Analysis

Code	Year	Details
ISO / CwD 13720		Enumeration of Pseudomonas spp. in meat and meat products
ISO 13493:	1998	Determination of chloramphenicol content- method using liquid chromatography in meat and meat products
ISO 13496:	2000	Detection of colouring agents - Method using thin layer chromatography in meat and meat products
ISO 13730:	1996	Determination of total phosphorus content -spectrometric method in meat and meat products
ISO 13965:	1998	Determination of starch and glucose contents -enzymatic method in meat and meat products

contd...

Code	Year	Details
ISO 1442:	1997	Determination of moisture content in meat and meat products
ISO 1443:	1973	Determination of total fat content in meat and meat products
ISO 1444:	1996	Determination of free fat content in meat and meat products
ISO 1841-1:	1996	Determination of chloride content in meat and meat products - Part 1: Volhard method
ISO 1841-2:	1996	Determination of chloride content in meat and meat products - Part 2: Potentiometric method
ISO 2294:	1974	Determination of total phosphorus content in meat and meat products
ISO 2917:	1999	Measurement of pH in meat and meat products
ISO 2918:	1975	Determination of nitrite content in meat and meat products
ISO 3091:	1975	Determination of nitrite content in meat and meat products
ISO 3496:	1994	Determination of hydroxyproline content in meat and meat products
ISO 4134:	1999	Determination of L- (+) glutamic acid content in meat and meat products
ISO 5553:	1980	Detection of polyphosphates in meat and meat products
ISO 5554:	1978	Determination of starch content in meat and meat products

contd...

Code	Year	Details
ISO 936:	1998	Determination of total ash in meat and meat products
ISO 937:	1978	Determination of nitrogen content in meat and meat products
ISO 1026:	1982	Fruit and vegetable products -- Determination of dry matter content by drying under reduced pressure and of water content by azeotropic distillation
ISO 10272-1:	2006	Microbiology of food and animal feeding stuffs -- Horizontal method for detection and enumeration of Campylobacter spp. -- Part 1: Detection method
ISO 10273:	2003	Microbiology of food and animal feeding stuffs -- Horizontal method for the detection of presumptive pathogenic Yersinia enterocolitica
ISO 10991:	2009	Micro process engineering -- Vocabulary
ISO 11290-1:	1996	Microbiology of food and animal feeding stuffs -- Horizontal method for the detection and enumeration of Listeria monocytogenes -- Part 1: Detection method
ISO 11290-2:	1998	Microbiology of food and animal feeding stuffs -- Horizontal method for the detection and enumeration of Listeria monocytogenes -- Part 2: Enumeration method

contd...

Code	Year	Details
ISO 13302:	2003	Sensory analysis -- Methods for assessing modifications to the flavour of foodstuffs due to packaging
ISO 14617-8:	2002	Graphical symbols for diagrams -- Part 8: Valves and dampers
ISO 14673-3:	2004	(IDF 189-3: 2004) Milk and milk products -- Determination of nitrate and nitrite contents -- Part 3: Method using cadmium reduction and flow injection analysis with in-line dialysis (Routine method)
ISO 15213:	2003	Microbiology of food and animal feeding stuffs -- Horizontal method for the enumeration of sulfite-reducing bacteria growing under anaerobic conditions
ISO 15214:	1998	Microbiology of food and animal feeding stuffs -- Horizontal method for the enumeration of mesophilic lactic acid bacteria -- Colony-count technique at 30 degrees C
ISO 16140:	2003	Microbiology of food and animal feeding stuffs -- Protocol for the validation of alternative methods
ISO 16634-1:	2008	Food products - Determination of the total nitrogen content by combustion according to the Dumas principle and calculation of the crude protein content -- Part 1: Oilseeds and animal feeding stuffs

contd...

Code	Year	Details
ISO 16649-1:	2001	Microbiology of food and animal feeding stuffs- Horizontal method for the enumeration of beta-glucuronidase-positive E. coli- Part 1: Colony-count technique at 44°C using membranes and 5-bromo-4-chloro-3-indolyl beta-D-glucuronide
ISO 16649-2:	2001	Microbiology of food and animal feeding stuffs -- Horizontal method for the enumeration of beta-glucuronidase-positive Escherichia coli -- Part 2: Colony-count technique at 44°C using 5-bromo-4-chloro-3-indolyl beta-D-glucuronide
ISO 16654:	2001	Microbiology of food and animal feeding stuffs -- Horizontal method for the detection of Escherichia coli *O157*
ISO 17410:	2001	Microbiology of food and animal feeding stuffs -- Horizontal method for the enumeration of psychrotrophic microorganisms
ISO 17604:	2003	Microbiology of food and animal feeding stuffs -- Carcass sampling for microbiological analysis
ISO 1842:	1991	Fruit and vegetable products -- Determination of pH
ISO 18593:	2004	Microbiology of food and animal feeding stuffs -- Horizontal methods for sampling techniques from surfaces using contact plates and swabs

contd...

Code	Year	Details
ISO 1871:	2009	Food and feed products -- General guidelines for the determination of nitrogen by the Kjeldahl method
ISO 2037:	1992	Stainless steel tubes for the food industry
ISO 20837:	2006	Microbiology of food and animal feeding stuffs -- Polymerase chain reaction (PCR) for the detection of food-borne pathogens -- Requirements for sample preparation for qualitative detection
ISO 20838:	2006	Microbiology of food and animal feeding stuffs -- Polymerase chain reaction (PCR) for the detection of food-borne pathogens -- Requirements for amplification and detection for qualitative methods
ISO 21469:	2006	Safety of machinery -- Lubricants with incidental product contact -- Hygiene requirements
ISO 21527-1:	2008	Microbiology of food and animal feeding stuffs -- Horizontal method for the enumeration of yeasts and moulds -- Part 1: Colony count technique in products with water activity greater than 0,95
ISO 21527-2:	2008	Microbiology of food and animal feeding stuffs -- Horizontal method for the enumeration of yeasts and moulds -- Part 2: Colony count technique in products with water activity less than or equal to 0,95

contd...

Code	Year	Details
ISO 21528-1:	2004	Microbiology of food and animal feeding stuffs -- Horizontal methods for the detection and enumeration of Enterobacteriaceae -- Part 1: Detection and enumeration by MPN technique with pre-enrichment
ISO 21528-2:	2004	Microbiology of food and animal feeding stuffs -- Horizontal methods for the detection and enumeration of Enterobacteriaceae -- Part 2: Colony-count method
ISO 21567:	2004	Microbiology of food and animal feeding stuffs -- Horizontal method for the detection of Shigella spp.
ISO 21569:	2005	Foodstuffs -- Methods of analysis for the detection of genetically modified organisms and derived products -- Qualitative nucleic acid based methods
ISO 21570:	2005	Foodstuffs -- Methods of analysis for the detection of genetically modified organisms and derived products -- Quantitative nucleic acid based methods
ISO 21571:	2005	Foodstuffs -- Methods of analysis for the detection of genetically modified organisms and derived products -- Nucleic acid extraction
ISO 2172:	1983	Fruit juice -- Determination of soluble solids content -- Pycnometric method

contd...

Code	Year	Details
ISO 21807:	2004	Microbiology of food and animal feeding stuffs -- Determination of water activity
ISO 21871:	2006	Microbiology of food and animal feeding stuffs -- Horizontal method for the determination of low numbers of presumptive Bacillus cereus -- Most probable number technique and detection method
ISO 22000:	2005	Food safety management systems -- Requirements for any organization in the food chain
ISO 22005:	2007	Traceability in the feed and food chain -- General principles and basic requirements for system design and implementation
ISO 22174:	2005	Microbiology of food and animal feeding stuffs -- Polymerase chain reaction (PCR) for the detection of food-borne pathogens -- General requirements and definitions
ISO 24276:	2006	Foodstuffs -- Methods of analysis for the detection of genetically modified organisms and derived products -- General requirements and definitions
ISO 2447:	1998	Fruit and vegetable products -- Determination of tin content
ISO 2448:	1998	Fruit and vegetable products -- Determination of ethanol content

contd...

Code	Year	Details
ISO 26642:	2010	Food products -- Determination of the glycaemic index (GI) and recommendation for food classification
ISO 28198:	2009	Vegetable fats and oils -- Determination of toluene insoluble matter
ISO 2851:	1993	Stainless steel bends and tees for the food industry
ISO 2852:	1993	Stainless steel clamp pipe couplings for the food industry
ISO 2853:	1993	Stainless steel threaded couplings for the food industry
ISO 30000:	2009	Ships and marine technology -- Ship recycling management systems -- Specifications for management systems for safe and environmentally sound ship recycling facilities
ISO 4531-1:	1998	Vitreous and porcelain enamels -- Release of lead and cadmium from enamelled ware in contact with food -- Part 1: Method of test
ISO 4531-2:	1998	Vitreous and porcelain enamels -- Release of lead and cadmium from enamelled ware in contact with food -- Part 2: Permissible limits
ISO 4831:	2006	Microbiology of food and animal feeding stuffs -- Horizontal method for the detection and enumeration of coliforms -- Most probable number technique

contd...

Code	Year	Details
ISO 4832:	2006	Microbiology of food and animal feeding stuffs -- Horizontal method for the enumeration of coliforms -- Colony-count technique
ISO 4833:	2003	Microbiology of food and animal feeding stuffs -- Horizontal method for the enumeration of microorganisms -- Colony-count technique at 30 degrees C
ISO 5498:	1981	Agricultural food products -- Determination of crude fibre content -- General method
ISO 5523:	1981	Liquid fruit and vegetable products -- Determination of sulphur dioxide content (Routine method)
ISO 5526:	1986	Cereals, pulses and other food grains -- Nomenclature
ISO 5667-21:	2010	Water quality -- Sampling -- Part 21: Guidance on sampling of drinking water distributed by tankers or means other than distribution pipes
ISO 5667-5:	2006	Water quality -- Sampling -- Part 5: Guidance on sampling of drinking water from treatment works and piped distribution systems
ISO 6486-1:	1999	Ceramic ware, glass-ceramic ware and glass dinnerware in contact with food -- Release of lead and cadmium -- Part 1: Test method

contd...

Code	Year	Details
ISO 6486-2:	1999	Ceramic ware, glass-ceramic ware and glass dinnerware in contact with food -- Release of lead and cadmium -- Part 2: Permissible limits
ISO 6541:	1981	Agricultural food products -- Determination of crude fibre content -- Modified Scharrer method
ISO 6560:	1983	Fruit and vegetable products -- Determination of benzoic acid content (benzoic acid contents greater than 200 mg per litre or per kilogram) -- Molecular absorption spectrometric method
ISO 6579:	2002	Microbiology of food and animal feeding stuffs -- Horizontal method for the detection of Salmonella spp.
ISO 6636-3:	1983	Fruit and vegetable products -- Determination of zinc content -- Part 3: Dithizone spectrometric method
ISO 6638-1:	1985	Fruit and vegetable products -- Determination of formic acid content -- Part 1: Gravimetric method
ISO 6638-2:	1984	Fruit and vegetable products -- Determination of formic acid content -- Part 2: Routine method

contd...

Code	Year	Details
ISO 6887-1:	1999	Microbiology of food and animal feeding stuffs -- Preparation of test samples, initial suspension and decimal dilutions for microbiological examination -- Part 1: General rules for the preparation of the initial suspension and decimal dilutions
ISO 6887-2:	2003	Microbiology of food and animal feeding stuffs -- Preparation of test samples, initial suspension and decimal dilutions for microbiological examination -- Part 2: Specific rules for the preparation of meat and meat products
ISO 6887-3:	2003	Microbiology of food and animal feeding stuffs -- Preparation of test samples, initial suspension and decimal dilutions for microbiological examination -- Part 3: Specific rules for the preparation of fish and fishery products
ISO 6887-4:	2003	Microbiology of food/animal feeding stuffs- Preparation of samples, initial suspension and decimal dilutions for micro. exam.- Part 4: Specific rules for the prep. of products other than milk/milk products, meat/meat products, fish/ fish products
ISO 6887-5:	2010	Microbiology of food and animal feeding stuffs -- Preparation of test samples, initial suspension and decimal dilutions for micro. exam.- Part 5: Specific rules for the preparation of milk and milk products

contd...

Code	Year	Details
ISO 6888-1:	1999	Microbiology of food and animal feeding stuffs -- Horizontal method for the enumeration of coagulase-positive staphylococci (Staphylococcus aureus and other species) -- Part 1: Technique using Baird-Parker agar medium
ISO 6888-2:	1999	Microbiology of food and animal feeding stuffs -- Horizontal method for the enumeration of coagulase-positive staphylococci (Staphylococcus aureus and other species) -- Part 2: Technique using rabbit plasma fibrinogen agar medium
ISO 6888-3:	2003	Microbiology of food and animal feeding stuffs -- Horizontal method for the enumeration of coagulase-positive staphylococci (Staphylococcus aureus and other species) -- Part 3: Detection and MPN technique for low numbers
ISO 7002:	1986	Agricultural food products -- Layout for a standard method of sampling from a lot
ISO 7086-1:	2000	Glass hollowware in contact with food -- Release of lead and cadmium -- Part 1: Test method
ISO 7086-2:	2000	Glass hollowware in contact with food -- Release of lead and cadmium -- Part 2: Permissible limits

contd...

Code	Year	Details
ISO 7218:	2007	Microbiology of food and animal feeding stuffs -- General requirements and guidance for microbiological examinations
ISO 7251:	2005	Microbiology of food and animal feeding stuffs -- Horizontal method for the detection and enumeration of presumptive Escherichia coli -- Most probable number technique
ISO 7466:	1986	Fruit and vegetable products -- Determination of 5-hydroxymethylfurfural (5-HMF) content
ISO 750:	1998	Fruit and vegetable products -- Determination of titratable acidity
ISO 751:	1998	Fruit and vegetable products -- Determination of water-insoluble solids
ISO 7700-1:	2008	Food products -- Checking the performance of moisture meters in use -- Part 1: Moisture meters for cereals
ISO 7932:	2004	Microbiology of food and animal feeding stuffs -- Horizontal method for the enumeration of presumptive Bacillus cereus -- Colony-count technique at 30 degrees C
ISO 7937:	2004	Microbiology of food and animal feeding stuffs -- Horizontal method for the enumeration of Clostridium perfringens -- Colony-count technique

contd...

Code	Year	Details
ISO 8086:	2004	(IDF 121: 2004) Dairy plant -- Hygiene conditions -- General guidance on inspection and sampling procedures
ISO 8128-1:	1993	Apple juice, apple juice concentrates and drinks containing apple juice -- Determination of patulin content -- Part 1: Method using high-performance liquid chromatography
ISO 8128-2:	1993	Apple juice, apple juice concentrates and drinks containing apple juice -- Determination of patulin content -- Part 2: Method using thin-layer chromatography
ISO 8156:	2005	(IDF 129: 2005) Dried milk and dried milk products -- Determination of insolubility index
ISO 8262-1:	2005	(IDF 124-1: 2005) Milk products and milk-based foods -- Determination of fat content by the Weibull-Berntrop gravimetric method (Reference method) -- Part 1: Infant foods
ISO 8262-2:	2005	(IDF 124-2: 2005) Milk products and milk-based foods -- Determination of fat content by the Weibull-Berntrop gravimetric method (Reference method) -- Part 2: Edible ices and ice-mixes

contd...

Code	Year	Details
ISO 8262-3:	2005	(IDF 124-3: 2005) Milk products and milk-based foods -- Determination of fat content by the Weibull-Berntrop gravimetric method (Reference method) -- Part 3: Special cases
ISO 8391-1:	1986	Ceramic cookware in contact with food -- Release of lead and cadmium -- Part 1: Method of test
ISO 8391-2:	1986	Ceramic cookware in contact with food -- Release of lead and cadmium -- Part 2: Permissible limits
ISO 8442-1:	1997	Materials and articles in contact with foodstuffs -- Cutlery and table holloware -- Part 1: Requirements for cutlery for the preparation of food
ISO 8442-5:	2004	Materials and articles in contact with foodstuffs -- Cutlery and table holloware -- Part 5: Specification for sharpness and edge retention test of cutlery
ISO 8967:	2005	(IDF 134: 2005) Dried milk and dried milk products -- Determination of bulk density
ISO 8381:	2008	(IDF 123: 2008) Milk-based infant foods -- Determination of fat content -- Gravimetric method (Reference method)
ISO 13580:	2005	(IDF 151: 2005) Yogurt -- Determination of total solids content (Reference method)

contd...

Code	Year	Details
ISO 17240:	2004	Fruit and vegetable products -- Determination of tin content -- Method using flame atomic absorption spectrometry
ISO 21527-1:	2008	Microbiology of food and animal feeding stuffs -- Horizontal method for the enumeration of yeasts and moulds -- Part 1: Colony count technique in products with water activity greater than 0,95
ISO 2173:	2003	Fruit and vegetable products -- Determination of soluble solids -- Refractometric method
ISO 22006:	2009	Quality management systems -- Guidelines for the application of ISO 9001:2008 to crop production
ISO 22855:	2008	Fruit and vegetable products-Determination of benzoic acid and sorbic acid concentrations - High performance liquid chromatography method
ISO 22866:	2005	Equipment for crop protection -- Methods for field measurement of spray drift
ISO 23393:	2006	Pomegranate fruit -- Specification and test methods
ISO 762:	2003	Fruit and vegetable products -- Determination of mineral impurities content
ISO 763:	2003	Fruit and vegetable products -- Determination of ash insoluble in hydrochloric acid

contd...

Code	Year	Details
ISO/ASTM 51204	2004	Practice for dosimetry in gamma irradiation facilities for food processing
ISO/ASTM 51431:	2005	Practice for dosimetry in electron beam and X-ray (bremsstrahlung) irradiation facilities for food processing
ISO/ASTM 51900:	2009	Guide for dosimetry in radiation research on food and agricultural products
ISO/TS 10272-2:	2006	Microbiology of food and animal feeding stuffs -- Horizontal method for detection and enumeration of Campylobacter spp. -- Part 2: Colony-count technique
ISO/TS 10272-3:	2010	Microbiology of food and animal feeding stuffs -- Horizontal method for detection and enumeration of Campylobacter spp. -- Part 3: Semi-quantitative method
ISO/TS 11133-1:	2009	Microbiology of food and animal feeding stuffs -- Guidelines on preparation and production of culture media -- Part 1: General guidelines on quality assurance for the preparation of culture media in the laboratory
ISO/TS 11133-2:	2003	Microbiology of food and animal feeding stuffs -- Guidelines on preparation and production of culture media -- Part 2: Practical guidelines on performance testing of culture media

contd...

Code	Year	Details
ISO/TS 16634-2:	2009	Food products - Determination of the total nitrogen content by combustion according to the Dumas principle and calculation of the crude protein content -- Part 2: Cereals, pulses and milled cereal products
ISO/TS 16649-3:	2005	Microbiology of food and animal feeding stuffs -- Horizontal method for the enumeration of beta-glucuronidase-positive Escherichia coli -- Part 3: Most probable number technique using 5-bromo-4-chloro-3-indolyl-beta-D-glucuronide
ISO/TS 19036:	2006	Microbiology of food and animal feeding stuffs -- Guidelines for the estimation of measurement uncertainty for quantitative determinations
ISO/TS 20836:	2005	Microbiology of food and animal feeding stuffs -- Polymerase chain reaction (PCR) for the detection of food-borne pathogens -- Performance testing for thermal cyclers
ISO/TS 21872-1:	2007	Microbiology of food and animal feeding stuffs -- Horizontal method for the detection of potentially enteropathogenic Vibrio spp. -- Part 1: Detection of Vibrio parahaemolyticus and Vibrio cholerae

contd...

Code	Year	Details
ISO/TS 21872-2:	2007	Microbiology of food and animal feeding stuffs -- Horizontal method for the detection of potentially enteropathogenic Vibrio spp. -- Part 2: Detection of species other than Vibrio parahaemolyticus and Vibrio cholerae
ISO/TS 22002-1:	2009	Prerequisite programmes on food safety -- Part 1: Food manufacturing
ISO/TS 22003:	2007	Food safety management systems -- Requirements for bodies providing audit and certification of food safety management systems
ISO/TS 22004:	2005	Food safety management systems -- Guidance on the application of ISO 22000:2005
ISO/TS 22117:	2010	Microbiology of food and animal feeding stuffs -- Specific requirements and guidance for proficiency testing by interlaboratory comparison

Appendix-XVIII

Light microscope v/s Electron Microsscope

Feature	Electron Microscope	Light Microscope
Highest practical magnification	Over 100,000	About 1,000-1,500
Best resolution*	0.5 nm	0.2μm
Radiation source	Electron beam	Visible light
Medium of travel	High vacuum	Air
Type of lens	Electromagnet	Glass
Source of contrast	Scattering of electrons	Differential light absorption
Focusing mechanism	Adjust current to the magnetic lens	Adjust lens position mechanically
Method of changing magnification	Adjust current to the magnetic lens	Switch the objective lens or eyepiece
Specimen mount	Metal grid (usually copper)	Glass slide

*The resolution limit of a human eye = 0.2 mm

Appendix-XIX

Major food borne Bacteria

Bacteria	Habitate	Source
Campylobacter	Intestinal tracts of animals, raw milk, untreated water, and sewage.	Contaminated water, raw milk, and raw or undercooked meat, poultry, or shellfish.
Botulism	Widely distributed in soil, water, on plants, and intestinal tracts of animals. Grows only in little or no oxygen.	Bacteria produce a toxin that causes illness. Improperly canned foods, garlic in oil, vacuum-packed and tightly wrapped food.
Clostridium	Soil, dust, sewage, and intestinal tracts of animals. Grows only in little or no oxygen.	Food left for long periods in steam tables or at room temperature. Bacteria destroyed by cooking, but some toxin-producing spores may survive.

Bacteria	**Habitate**	**Source**
E. coli	Intestinal tracts of some mammals, raw milk and unchlorinated water.	Contaminated water, raw milk, raw or rare ground beef, unpasteurized apple juice or cider, uncooked fruits and vegetables; person-to-person.
Listeria	Intestinal tracts of humans and animals, milk, soil, leaf vegetables; can grow slowly at refrigerator temperatures.	Ready-to-eat foods such as hot dogs, luncheon meats, cold cuts, fermented or dry sausage, and other deli-style meat and poultry, soft cheeses and unpasteurized milk.
Salmonella (over 2300 types)	Intestinal tracts and feces of animals and sometimes in eggs.	Raw or undercooked eggs, poultry, and meat; raw milk and dairy products; seafood, and food handlers.
Shigella (over 30 types)	Human intestinal tract; rarely found in other animals.	Person-to-person by fecal-oral route; fecal contamination of food and water. Most outbreaks result from food, especially salads, prepared and handled by workers using poor personal hygiene.
Staphylococcus	On humans (skin, infected cuts, pimples, noses, and throats).	Person-to-person through food from improper food handling. Multiply rapidly at room temperature to produce a toxin that causes illness.

Appendix-XX

Measurement Equivalents

1/5 teaspoon	=	1	milliliter
1 teaspoon	=	5	ml
1 tablespoon	=	15	ml
1/5 cup	=	50	ml
1 cup	=	240	ml
2 cups (1 pint)	=	470	ml
4 cups (1 quart)	=	.95	liter
4 quarts (1 gal.)	=	3.8	liters
1 fluid oz.	=	30	milliters
1 fluid oz.	=	28	grams
1 pound	=	454	grams
1 militers	=	1/5	teaspoon
5 ml	=	1	teaspoon
15 ml	=	1	tablespoon
34 ml	=	1	fluid oz.
100 ml	=	3.4	fluid oz.

240 ml	=	1	cup
1 liter	=	34	fluid oz.
1 liter	=	4.2	cups
1 liter	=	2.1	pints
1 liter	=	1.06	quarts
1 liter	=	.26	gallon
1 gram	=	.035	ounce
100 grams	=	3.5	ounces
500 grams	=	1.10	pounds
1 kilogram	=	2.205	pounds
1 kilogram	=	35	oz.
16 tablesppns	=	1	cup
12 tablespoons	=	3/4	cup
10 tablespoons + 2 teaspoons	=	2/3	cup
8 tablespoons	=	1/2	cup
6 tablespoons	=	3/8	cup
5 tablespoons + 1 teaspoon	=	1/3	cup
4 tablespoons	=	1/4	cup
2 tablespoons	=	1/8	cup
2 tablespoons + 2 teaspoons	=	1/6	cup
1 tablespoon	=	1/16	cup
2 cups	=	1	pint
2 pints	=	1	quart
3 teaspoons	=	1	tablespoon
48 teaspoons	=	1	cup

Appendix-XXI

Methods Used to Study Microorganisms in Different Environments

Characteristic	Evaluated Technique Employed or Property Measured	Environment				
		Marine	Freshwater	Sewage	Soil	Food
Nutrients						
	Chemical analysis (e.g., C, N, P)	++	++	++	++	++
	COD (chemical oxygen demand)	-	+	++	-	-
	BOD (biochemical oxygen demand)	-	+	++	-	-
Microbial biomass						
	Photosynthetic pigments	++	++	-	-	-
	Filtration and dry weight	++	++	+	-	+
	Measurement of chemical constituents	++	++	++	++	++

Characteristic	Evaluated Technique Employed or Property Measured	Environment				
		Marine	Freshwater	Sewage	Soil	Food
	(e.g., ATP, muramic acid, polybetahydroxybutyric [PHB] acid, lipopolysaccharides)					
	Microscopy and biovolume conversion to biomass using conversion factors	++	++	+	++	++
	Fumigation incubation/extraction	-	-	+	++	-
	Glucose-amended respiration	-	-	+	++	
Microbial numbers/types						
	Microscopic procedures-epifluorescence/ phase contrast	++	++	++	++	+
	Flow cytometry	++	++	+	-	-
	Immersion/insertion slides and films	+	++	+	++	-
	Viable enumeration procedures (culture, microscopy)	++	++	++	++	++
	Direct microorganism isolation	++	++	++	++	++
	Thin sections of samples	-	-	-	++	++

Characteristic	Evaluated Technique Employed or Property Measured	Environment				
		Marine	Freshwater	Sewage	Soil	Food
	Scanning electron microscopy	++	++	++	++	++
	Direct DNA extraction and analysis	++	++	++	++	++
	Polymerase chain reaction (PCR) with species-specific primers	++	++	++	++	++
	DNA probes and hybridization techniques for DNA and RNA	++	++	++	++	++
	In situ PCR with 16S or 18S primers	++	++	++	++	++
	Micromanipulation and single-cell PCR/ phylogenetic analysis	++	++	++	++	++
Microbial viability and turnover						
	Nalidixic acid and microscopic observation (inhibitor prevents cell division, resulting in elongated active cells); direct viable counts	++	++	++	++	++
	Bioluminescence assessment (green fluorescent protein)	+	+	+	+	+
	Stable and radioactive isotope studies	++	++	+	++	++

Characteristic	Evaluated Technique Employed or Property Measured	Environment				
		Marine	Freshwater	Sewage	Soil	Food
Microbial activity						
	Microscopy with reducible dyes	++	++	++	++	++
	Autoradiography	++	++	++	++	++
	Enzyme activity assays	+	+	++	++	+
	Microcalorimetry	-	-	+	+	+
	RNA-based function probes	++	++	++	++	++
	Gas exchange (O2, CO2, N2, CH4)	++	++	++	++	-
	Assessment of fungal and bacterial contributions by use of selective antibiotic inhibition	+	+	-	++	-
	Substrate utilization rate	++	++	++	++	+
	Fluorescent substrate hydrolysis	++	++	++	++	+
Community structure						
	Microscopic analyses of diversity	++	++	++	++	+
	Physiological diversity of microbial isolates	++	++	++	++	+

Characteristic	Evaluated Technique Employed or Property Measured	Environment				
		Marine	Freshwater	Sewage	Soil	Food
	16S or 18S ribosomal RNA analysis (including SSCP [single-strand conformation polymorphism], DGGE [denaturing gradiant gel electrophoresis], TGGE [temperature gradient gel electrophoresis], etc.)	++	++	++	++	+
	DNA amplification fingerprinting (DAF)	++	++	++	++	+

[a]Major uses (++); minor uses (+); minimal or no use (-)

Appendix-XXII

Microorganisms: G+C Contents

Organism	percent G+C	Organism	percent G+C
Bacteria			
Actinomyces	*59-73*	*Anabaena*	*38-44*
Bacillus	*32-62*	*Bacteroides*	*28-61*
Bdellovibrio	*33-52*	*Caulobacter*	*63-67*
Chlamydia	*41-44*	*Chlorobium*	*49-58*
Chromatium	*48-70*	*Clostridium*	*21-54*
Cytophaga	*33-42*	*Deinococcus*	*62-70*
Escherichia	*48-52*	*Halobacterium*	*66-68*
Hyphomicrobium	*59-67*	*Methanobacterium*	*32-50*
Micrococcus	*64-75*	*Mycobacterium*	*62-70*
Mycoplasma	*23-40*	*Myxococcus*	*68-71*
Neisseria	*47-54*	*itrobacter*	*60-62*
Oscillatoria	*40-50*	*Prochloron*	*41*
Proteus	*38-41*	*Pseudomonas*	*58-70*
Rhodospirillum	*62-66*	*Rickettsia*	*29-33*

Organism	percent G+C	Organism	percent G+C
Salmonella	50-53	*Spirillum*	38
Spirochaeta	51-65	*Staphylococcus*	30-38
Streptococcus	33-44	*Streptomyces*	69-73
Sulfolobus	31-37	*Thermoplasma*	46
Thiobacillus	52-68	*Treponema*	25-54
Algae			
Acetabularia mediterranea	37-53	*Chlamydomonas*	60-68
Chlorella	43-79	*Cyclotella cryptica*	41
Euglena gracilis	46-55	*Nitella*	49
Nitzschia angularis	47	*Ochromonas danica*	48
Peridinium triquetrum	53	*Scenedesmus*	52-64
Spirogyra	39	*Volvox carteri*	50
Protozoa			
Acanthamoeba castellanii	56-58	*Amoeba proteus*	66
Parameciumspp.	29-39	*Plasmodium berghei*	41
Stentor polymorphus	45	*Tetrahymena*	19-33
Trichomonas	29-34	*Trypanosoma*	45-59
Slime Molds			
Dictyostelium	22-25	*Lycogala*	42
Physarum polycephalum	38-42		

Organism	percent G+C	Organism	percent G+C
Fungi			
Agaricus bisporus	*44*	*Amanita muscaria*	*57*
Aspergillus niger emersonii	*52*	*Blastocladiella*	*66*
Candida albicans	*33-35*	*Claviceps purpurea*	*53*
Coprinus lagopus	*52-53*	*Fomes fraxineus*	*56*
Mucor rouxii	*38*	*Neurospora crassa*	*52-54*
Penicillium notatum	*52*	*olyporus palustris*	*56*
Rhizopus nigricans	*47*	*Saccharomyces cer.*	*36-42*

Appendix-XXIII

Mycotoxins

Disease[a]	Fungus	Mycotoxin	Contaminated Foodstuff	Animals Affected
Aflatoxicosis	Aspergillus flavus	Aflatoxins	Rice, corn, sorghum, cereals, peanuts, soybeans	Poultry, swine, cattle, sheep, dogs
Ergotism	Claviceps purpurea	Ergot alkaloids	Seedheads of many grasses, grains	Cattle, horses, swine, poultry
Mushroom poisoning	Amanita verna	Amanitins	Eaten from pastures	Cattle

Disease[a]	**Fungus**	**Mycotoxin**	**Contaminated Foodstuff**	**Animals Affected**
Poultry hemorrhagic syndrome	Aspergillus flavus and others	Aflatoxins	Toxic grains and meal	Chickens
Slobbers slaframine	Rhizoctonia	Alkaloid	Red clover	Sheep, cattle
Tall fescue toxicosis	Acremonium coenophialum (an endophytic fungus)	Ergot alkaloids	Endophyte-infected tall fescue plants	Cattle, horses

Appendix-XXIV

Properties of Microscope Objectives

Objective

Properties	Scanning	Low Power	High Power	Oil Immersion
Magnification	4×	10×	40-45×	90-100×
Numerical aperture	0.10	0.25	0.55-0.65	1.25-1.4
Approximate focal length (f)	40 mm	16 mm	4 mm	1.8-2.0 mm
Working distance	17-20 mm	4-8 mm	0.5-0.7 mm	0.1 mm
Approximate resolving power with light of 450 nm (blue light)	2.3 µm	0.9 µm		0.35 µm 0.18 µm

Appendix-XXV

Relative Density (Specific Gravity)

Water (sea average)	1.03	Nickel	8.6
Aluminum	2.56	Oil (linseed)	0.94
Antimony	6.70	Oil (olive)	0.92
Bismuth	9.80	Oil (petroleum)	0.76-0.86
Brass	8.40	Oil (turpentine)	0.87
Brick	2.1	Paraffin	0.86
Calcium	1.58	Platinum	21.5
Carbon (diamond)	3.4	Sand (dry)	1.42
Carbon (graphite)	2.3	Silicon	2.6
Carbon (charcoal)	1.8	Silver	10.57
Chromium	6.5	Slate	2.1-2.8
Clay	1.9	Sodium	0.97
Coal	1.36-1.4	Steel (mild)	7.87
Cobalt	8.6	Sulphur	2.07
Copper	8.77	Tin	7.3
Cork	0.24	Tungsten	19.1
Glass (crown)	2.5	Wood (ash)	0.75

Glass (flint)	3.5	Wood (beech)	0.7-0.8
Gold	19.3	Wood (ebony)	1.1-1.2
Iron (cast)	7.21	Wood (elm)	0.66
Iron (wrought)	7.78	Wood (lignum-vitae)	1.3
Lead	11.4	Wood (oak)	0.7-1.0
Magnesium	1.74	Wood (pine)	0.56
Manganese	8.0	Wood (teak)	0.8
Mercury	13.6	Zinc	7.0

Appendix-XXVI

Selected Germicides

Class	Concentration of Active Ingredient	Activity Level[a]
Gas		
Ethylene oxide	450-500 mg/liter[b]	H
Liquid		
Glutaraldehyde, aqueous	2%	H to I
Formaldehyde + alcohol	8 + 70%	H
Stabilized hydrogen peroxide	6-30%	H to I
Formaldehyde, aqueous	6-8%	H to I
Iodophors	750-5,000 mg/liter[c]	H to I
Iodophors	75-150 mg/liter[c]	I to L
Iodine + alcohol	0.5 + 70%	I
Chlorine compounds	0.1-0.5%[d]	I
Phenolic compounds, aqueous	0.5-3%	I to L
Iodine, aqueous	1%	I
Alcohols (ethyl, isopropyl)	70%	I

Class	Concentration of Active Ingredient	Activity Level[a]
Quaternary ammonium. comp	0.1-0.2% aqueous	L
Chlorhexidine	0.75-4%	L
Hexachlorophene	1-3%	L
Mercurial compounds	0.1-0.2%	L

[a]High-level (H) disinfectants destroy vegetative bacterial cells including M. tuberculosis, bacterial endospores, fungi, and viruses. Intermediate-level (I) disinfectants destroy all of the above except endospores. Low-level (L) agents kill bacterial vegetative cells except for M. tuberculosis, fungi, and medium-sized lipid-containing viruses (but not bacterial endospores or small, nonlipid viruses).

[b]In autoclave-type equipment at 55 to 60°C.

[c]Available iodine.

[d]Free chlorine.

Appendix-XXVII

Temperature Ranges for Microbial Growth

	Cardinal temperature °C		
	Minimum	**Optimum**	**Maximum**
Nonphotosynthetic Procaryotes			
Bacillus psychrophilus	-10	23-24	28-30
Micrococcus cryophilus	-4	10	24
Pseudomonas fluorescens	4	25-30	40
Staphylococcus aureus	6.5	30-37	46
Enterococcus faecalis	0	37	44
Escherichia coli	10	37	45
Neisseria gonorrhoeae	30	35-36	38
Thermoplasma acidophilum	45	59	62
Bacillus stearothermophilus	30	60-65	75
Thermus aquaticus	40	70-72	79

	Minimum	Optimum	Maximum
Sulfolobus acidocaldarius	60	80	85
Pyrococcus abyssi	67	96	102
Pyrodictium occultum	82	105	110
Pyrolobus fumarii	90	106	113
Photosynthetic Bacteria			
Rhodospirillum rubrum	-	30-35	-
Anabaena variabilis	-	35	-
Oscillatoria tenuis	-	-	45-47
Synechococcus eximius	70	79	84
Eucaryotic Algae			
Chlamydomonas nivalis	-36	0	4
Fragilaria sublinearis	-2 5	-6	8-9
Chlorella pyrenoidosa	-	25-26	29
Euglena gracilis	-	23	-
Skeletonema costatum	6	16-26	>28
Cyanidium caldarium	30-34	45-50	56
Fungi			
Candida scottii	0	4-15	15
Saccharomyces cerevisiae	1-3	28	40
Mucor pusillus	21-23	45-50	50-58
Protozoa			
Amoeba proteus	4-6	22	35
Naegleria fowleri	20-25	35	40
Trichomonas vaginalis	25	32-39	42
Paramecium caudatum	-	25	28-30

Appendix-XXVIII

Types of Microbes (Environmental Determinants)

Descriptive Term	Definition	Representative Microorganisms
Solute and Water Activity		
Osmotolerant	Able to grow over wide ranges of water activity or osmotic concentration	*Staphylococcus aureus*
Halophile	Requires high levels of sodium chloride, usually above about 0.2 M, to grow	*Halobacterium*
pH		
Acidophile	Growth optimum between pH 0 and 5.5	*Sulfolobus*, *Picrophilus*, *Acontium*
Neutrophile	Growth optimum between pH 5.5 and 8.0	*Escherichia*, *Euglena*, *Paramecium*

Descriptive Term	Definition	Representative Microorganisms
Alkalophile	Growth optimum between pH 8.5 and 11.5	*Bacillus alcalophilus*
Temperature		
Psychrophile	Grows well at 0°C and has an optimum growth temperature of 15°C or lower	*Bacillus psychrophilus*
Psychrotroph	Can grow at 0-7°C; has an optimum between 20 and 30°C and a maximum around 35°C	*Listeria monocytogenes*
Mesophile	Has growth optimum around 20-45°C	*Escherichia coli*
Thermophile	Can grow at 55°C or higher; optimum often between 55 and 65°C	*Bacillus stearothermophilus*
Hyperthermophile	Has an optimum between 80 and about 113°C	*Sulfolobus, Pyrococcus*
Oxygen Concentration		
Obligate aerobe	Completely dependent on atmospheric $O^2$for growth.	*Pseudomonas, Mycobacterium*

Descriptive Term	Definition	Representative Microorganisms
Facultative anaerobe	Does not require O_2 for growth, but grows better in its presence.	*Escherichia, Enterococcus*
Obligate anaerobe	Does not tolerate O_2 and dies in its presence.	*Clostridium, Bacteroides*
Microaerophile	Requires O_2 levels below 2-10% for growth and is damaged by atmospheric O_2 (20%).	*Campylobacter*
Pressure		
Barophilic	Growth more rapid at high hydrostatic. pressures	*Photobacterium profundum*

Appendix-XXIX

Units of Measure

SIZE
(Relative to a meter)

Symbol - Measure			Relative Length (meter)	Exponential Notation (meter)
km	-	kilometer	1000	10^3
m	-	meter	1	10^0
dm	-	decimeter	0.1	10^{-1}
cm	-	centimeter	0.01	10^{-2}
mm	-	millimeter	0.001	10^{-3}
μm	-	micrometer or micron	0.000001	10^{-6}
nm	-	nanometer	0.000000001	10^{-9}
Å	-	Angstrom	0.0000000001	10^{-10}

Light microscope level - micron; Electron microscope level - nanometer; Molecular measurements – Angstrom

WEIGHT
(Relative to a gram)

Symbol - Measure			Relative Weight (gram)	Exponential Notation (gram)
kg	-	kilogram	1000	10^{3}
g	-	gram	1	10^{0}
mg	-	milligram	0.001	10^{-3}
µg	-	microgram	0.000001	10^{-6}
ng	-	nanogram	0.000000001	10^{-9}
pg	-	picogram	0.000000000001	10^{-12}
fg	-	femtogram	0.000000000000001	10^{-15}

VOLUME
(Relative to a liter)

Symbol - Measure			Relative Volume (litre)	Exponential Notation (litre)
L	-	liter	1	10^{0}
dL	-	deciliter	0.1	10^{-1}
mL	-	milliliter	0.001	10^{-3}
µL	-	microliter	0.000001	10^{-6}
nL	-	nanoliter	0.000000001	10^{-9}
pL	-	picoliter	0.000000000001	10^{-12}
fg	-	femtoliter	0.000000000000001	10^{-15}

MOLE

Symbol - Measure		Relative measure (mole)	Exponential Notation (mole)
mol -	mole	1	10^{0}
dmol -	decimole	0.1	10^{-1}
mmol -	millimole	0.001	10^{-3}
µmol -	micromole	0.000001	10^{-6}
nmol -	nanomole	0.000000001	10^{-9}
pmol -	picomole	0.000000000001	10^{-12}
fmol -	femtomole	0.000000000000001	10^{-15}

A **mole** is equal to the gram molecular weight (or formula weight) of the solute.

Molarity (M) equals the number of moles of solute in 1 liter of solution.

A **molar solution** is one in which 1 litre of solution contains the number of grams equal to its molecular weight.
Example: Sodium Chloride (NaCl) has a formula (molecular) weight of 58.44 (22.990 for Na and 35.453 for Cl).To make up 1L of a 5M NaCl solution = 58.44 g/mol x 5 moles/liter x 1 liter = 292.2 g in 1 liter of solution.

Molality (m) is equal to the number of moles of solute in 1000 g of solvent. For dilute solutions in water (a ml of water equals 1.0 g), there is little or no practical difference between a molar solution and a molal solution. In contrast, a 2 m sucrose solution (density of 1.18 g/ml) equals a 1.4 M sucrose solution.

Bibliography

40 steps to a safer laboratory (*http://www.labsafety.org*) Access Date: 6 Nov 2010.

Basics for quality assurance for intermediate and peripheral laboratories (WHO publication) biosafetypage/section4.htm). ***Princeton University***. Access Date: 14 June 2010.

Catalogues of individual equipment in relation to standard operating procedures, precautions and maintenance, ***Various suppliers, Various publication date***.

Chemical Hygiene Plan. School of chemical Sciences, ***University of Illinois***, January 2010.

Chemical Management and Inventory System Manual. ***Ateneo de Manila University***. 2008

Environmental Health and Safety (*http://web.princeton.edu/sites/ehs/biosafety/*biosafetypage/section4.htm). ***Princeton University***. Access Date: 14 June 2010.

First Aid Manual, 8th Edition. St John Ambulance, St Andrews First Aid, British Red Cross, 2002

Frequently Asked Safety Questions (*http://www.flinnsci.com*) Access Date: 16 Oct 2010.

Hazardous and Toxic Substances. Occupational Safety and Health Administration (OSHA) (*http://www.osha.gov/SLTC/hazardoustoxicsubstances/*) Date Accessed: 17 Oct 2010.

WHO Guidelines on standard operating procedures for clinical chemistry, Sep 2000

WHO publications / internet sources/handouts.